T0135210

Coastal Ocean Observing

Frontispiece: Tropical Ocean Observing
Art courtesy of Mr. Mark Sabino, member of the CariCOOS Board of Directors

Jorge E. Corredor

Coastal Ocean Observing

Platforms, Sensors and Systems

Jorge E. Corredor
Department of Marine Sciences (retired)
University of Puerto Rico
Mayagüez, Puerto Rico

ISBN 978-3-030-08684-8 ISBN 978-3-319-78352-9 (eBook)
https://doi.org/10.1007/978-3-319-78352-9

Softcover re-print of the Hardcover 1st edition 2018

Printed on acid-free paper

This Springer imprint is published by the registered company Springer International Publishing AG part of Springer Nature.
The registered company address is: Gewerbestrasse 11, 6330 Cham, Switzerland

To past and present members of the Caribbean Coastal Ocean Observing System (CariCOOS) whose collective knowledge I hope to here accurately reflect.

To the United States Integrated Ocean Observing System (US IOOS) and its predecessor OCEAN.US, institutions that provided funding and guidance without which CariCOOS would have been impossible.

To members of the other ten fellow US IOOS Coastal Ocean Observing Systems who generously provided support, advice, and encouragement in the establishment of CariCOOS.

Preface

This book arises from material initially compiled for a practical graduate field course on oceanographic techniques taught at the University of Puerto Rico, Mayagüez Campus (UPRM), Department of Marine Sciences at La Parguera Puerto Rico over a period of 35 years. Oceanographic techniques encompass practices developed to observe ocean properties and obtain experimental samples, practices which are in large part *remote* since instrument deployment is performed mostly from vessels at sea. The course included the operation and maintenance of oceanographic probes provided with a variety of electronic and electro-optical sensors, and of an equally wide variety of instruments deployed for remote sampling of water, sediment, and plankton. Data collection and processing from instrument casts as well as handling and preservation of samples obtained with the various collection devices were included in the curriculum. Concurrent graduate courses on theoretical and practical aspects of chemical oceanography and marine pollution provided material regarding electrochemical sensors, their design, operation, and limitations.

In 1993, support became available for the implementation of an oceanic time series observing effort off the south coast of Puerto Rico. The Caribbean Time Series (CaTS) was occupied at monthly intervals aboard various oceanographic vessels through the year 2006. Vertical profiles of physical and biogeochemical water column features were obtained using instruments here described beginning with casts to 200 m depth and eventually reaching depths of 3000 m. The CaTS effort provided a seagoing laboratory for these courses and spurred periodic updates as new instruments and techniques became available.

Support for a Caribbean Coastal Ocean Observing System (CariCOOS) for Puerto Rico and the United States Virgin Islands (PR and USVI) within a nationwide Integrated United States Ocean Observing System (US IOOS) beginning around 2004 provided unprecedented autonomous observing capabilities. Through the efforts of researchers at the University of Puerto Rico and the University of the Virgin Islands, a multicomponent coastal ocean observing system was planned, designed, and implemented and is now operational. The reward has been a wealth of data-rich sources including instrumented buoys, coastal HF radar stations, and operational autonomous glider transects yielding in turn a wealth of data products in user-friendly formats serving the coastal maritime community.

The 11 independent but coordinated Integrated Coastal Ocean Observing Systems established in US coastal waters have filled a void of information that has already proven its values in daily use as well as in several high profile events and climate-driven emergencies. Through these efforts, sustained ocean observations now allow the publication of operational data products available through a wide variety of electronic means to stakeholders and the general public. It has now become possible for the mariner to know the measured winds and waves, the temperatures and currents at several sites throughout the coastal zone with data updated within the hour through internet and cellular network connectivity. Moreover, this data, coupled to numerical models guided by assimilation of real-time instrumental data, has allowed operational implementation of hindcasts, nowcasts, and short-term forecasts of ocean conditions at a scale commensurate to the needs of the coastal ocean stakeholder.

The shift in observing strategies to prominently include autonomous land- and sea-based electronic sensing systems prompted refocusing of the coursework at UPRM to include, in addition to expeditionary oceanographic techniques, the practice of coastal ocean observing using autonomous observing platforms. Happily, these courses spanned the period over which most of the modern suite of instruments and platforms became commercially available. Many of these were put to use and indeed are still operational as part of the current observing effort.

Active involvement of the author in the establishment of the Caribbean Regional Association for Coastal Ocean Observing (today a not-for-profit entity incorporated in the state of Puerto Rico under the name of CariCOOS) provided first-hand experience in planning and implementation of observing systems. During the development of CariCOOS, opportunity arose for implementing high frequency radar as a dual-use technology to detect and track vessels at sea while simultaneously tracking ocean surface currents over a wide swath of the coastal ocean. Numerous other opportunities arose for mutually beneficial collaboration with various academic, governmental, and private-sector organizations which are discussed below in greater detail. Concurrent service of the author on the Ocean Studies Board of the US National Academies of Science, Engineering, and Medicine provided expert views on many of the topics here discussed, particularly regarding ocean observing, numerical modeling, and carbon biogeochemistry.

The book is biased to the practice of ocean observing in the balmy tropics. Due apologies are extended to colleagues working in more rigorous climes and to the reader for the comparative void of information in this regard. Likewise, international readers outside the USA will find the book necessarily biased to the author's experience within this jurisdiction to which PR and USVI are subject.

Professor Aurelio Mercado offered invaluable assistance in clarifying and correcting concepts regarding numerical modeling. Drs. Roy Armstrong and Miguel Canals were extremely helpful in the graphic documentation of oceanographic instruments and platforms in current operational use. Most of the equipment here depicted is composed of working units. The challenges of the environment where they are deployed are readily apparent.

Mayagüez, Puerto Rico Jorge E. Corredor

The original version of this book was revised. A correction to this book can be found at https://doi.org/10.1007/978-3-319-78352-9_9

Contents

Chapter 1
Introduction to Coastal Ocean Observing

> *One major development of the past decade was the advancement of operational oceanography and, specifically, the implementation of ocean observing systems that encompass observations, models, and analysis to yield societally relevant oceanographic information in near real time.*
>
> *Edwards et al. (2015)*

Abstract Technology developments in the fields of electronic sensing, signal amplification, communications, and autonomous navigation have led to the design, manufacture and deployment of autonomous environmental sensors in distributed networks allowing monitoring of a number of environmental variables in near real time at multiple locations. Autonomous instrument-laden platforms plumb the ocean depths at unprecedented data rates, and active and passive electromagnetic sensing instruments aboard satellites in terrestrial orbit provide wide ranging synoptic views of ocean surface and subsurface features. Advances in autonomous remote sensing are contrasted to the historical practice of ocean observing aboard manned vessels. The nature and priorities of operational coastal observing systems are set forth emphasizing the timely release of data and data products tailored to provide *societally relevant* oceanographic information.

Keywords Electronic sensing · Distribute networks · Autonomous platforms · Data products

Modern electronic environmental sensors using recently developed materials can quantify states and process rates for numerous physical and biogeochemical variables. Parallel advances in integrated electronic circuitry allow the ability to collect data at unprecedented rate, accuracy, and precision. Data processing, integration

---

The original version of this chapter was revised. A correction to this chapter can be found at https://doi.org/10.1007/978-3-319-78352-9_9

J. E. Corredor, *Coastal Ocean Observing*,
https://doi.org/10.1007/978-3-319-78352-9_1

and telemetry, battery storage capacity, and electronic 3-D navigation have equally improved. Availability of novel primary transducers together with these advances has led to the development of robust, miniaturized, field deployable instruments. The advent of electro-optical devices has revolutionized the capability for detecting and measuring a wide range of chemical and biological variables and processes.

These advances have now reached the point of allowing sustained, widely distributed collection of environmental data by compact, autonomous instrument systems. Wide band dual communications allow remote operation of these networks with ever-increasing capabilities. Hart and Martinez (2006) define such integrated systems as *environmental sensor networks* where these capabilities are integrated into systems providing multilayered, data-dense views of spatial and temporal variability of environmental conditions.

In the field of ocean science, expeditionary oceanographic research aboard manned vessels provided an important testbed for the design and development of such instrument systems. Today, instruments recording temperature and salinity and other variables routinely operate at data sampling rates up to 24 Hz. Vertically operated profiling instruments known as CTDs (for conductivity (C), temperature (T) and depth (D)), descending at rates up to 60 $m.min^{-1}$ thus achieve sampling densities up to 24 data points per meter or 120,000 data points for a full ocean depth cast to 5000 m.

Instrumental Data: Then and Now

Fifty years ago a vertical hydrographic wire cast from a ship sampling to full ocean depth would have sampled 24 data points using reversing mercury thermometers for temperature measurement mounted on bottle samplers for subsequent laboratory salinity and oxygen analyses making a total of 86 data records including depth, derived from temperature anomalies of protected versus non-protected thermometer pairs. Bottles were affixed sequentially to a weighted wire rope and then tripped by means of bronze *messengers* (weights sequentially traveling down the wire rope) to invert the thermometers and simultaneously trip the bottle to capture a water sample. Paired thermometers were read at sea (through a handheld magnifying glass) upon retrieval of the array and salinity was determined with bench salinometers in the laboratory. Dissolved oxygen and a few other variables were measured in the laboratory using wet chemical techniques. The same cast today, performed with sensor-based electronic instrumentation, obtains 5000 times more coupled depth, temperature, salinity and oxygen data points with real-time graphical representation, electronic readout, and digital data recording. Data density may be increased severalfold by addition of various optical, bio-optical, and opto-chemical sensor devices to the instrument package.

CTD and shipboard flow-through systems have evolved into multiparameter data acquisition systems incorporating a variety of optical, chemical, and biophysical sensors. Many current profiling instrument packages accommodate modular sensors interchangeable in the field as may be required in addition to the traditional pressure, temperature, and conductivity sensors. Many versions of these instruments, first developed for cable deployment, are now employed in shipboard or shore-based infrastructure using pumped flow-through sensor systems. Flow-through sensors, vertical profiling and sampling systems, and towed vehicle-mounted instruments have vastly multiplied the data-gathering capability of research vessels.

Despite such advances, sustained coastal ocean observing programs involving periodic occupation of established stations and/or transect lines remain rare due to the expense of operating manned vessels at sea which can range well into the tens of thousands of dollars per day. Fisheries surveys and fishery-related data gathering remain the exception although some such observing mission are now being performed by autonomous surface and underwater vessels equipped with acoustic fish sensing instrumentation. Remarkable among such sustained, manned vessel-based efforts is the CALCOFI Survey, arising from the preceding California Cooperative Sardine Research Program dating to 1949. CALCOFI arose in response to the collapse of the California sardine fishery. Today, the survey incorporates 60 core stations along 11 transect lines normal to the California coast. While a large part of the effort is devoted to fish stock assessment through various means, hydrographic profiles including physical, chemical, and biological variables are secured at all stations. CTD/rosette casts provide instrumental profiles plus bottle samples. Variables that require calibration samples, or those for which electronic transducers do not exist, are measured from bottle samples. Process rates such as primary biological production (photosynthetic rate) are also measured.

Oceanic in character and research-driven in practice the Bermuda Atlantic Time Series (BATS) and the Hawaii Ocean Time Series (HOT) also deserve mention. Large, research ships equipped with sophisticated sampling systems and shipboard laboratories are dedicated to long-term documentation of biogeochemical ocean properties and processes and their response to climate forcing. Together, these efforts, occupying single stations at monthly intervals have provided irrefutable evidence for a long-term ocean warming trend and have demonstrated strong covariance of ocean pH decrease with the increasing atmospheric CO_2 load (Dore et al. 2009).

Today, expeditionary oceanography is increasingly being supplemented, and replaced in some cases, by instrumental observations using a variety of autonomous stationary and mobile platforms equipped with dedicated suits of advanced sensors coupled to electronic navigational, computational, and telemetric packages. Coastal ocean observing systems, in particular, have burgeoned in recent years spurred by these advances and responding to the growing needs of a wide range of stakeholders operating in this domain.

Ocean observatories, of great value to oceanographic research and in some ways akin to the astronomical observatories, are primarily concerned with the advancement of science. Ocean observing systems, on the other hand, while invariably useful to science, are primarily dedicated to serving stakeholder needs. Stakeholders in the commercial, conservation, recreational, regulatory, security, and scientific fields increasingly rely on observing system data products, nowcasts, and forecasts for operational planning and execution. Such packaged data products are now supplied by government-supported integrated coastal ocean observing systems (such as those forming part of the United States *Integrated Ocean Observing System* IOOS) as well as by commercial enterprises. These developments have brought us to the dawn of an era of truly operational autonomous ocean observing.

This book is focused on the practice of operational coastal ocean observing providing data and data products useful to the stakeholder. The book describes the wide available range of electromechanical, electrochemical, electro-optical, and electro-acoustic sensor systems at the heart of current field-deployable ocean observing instruments. Their principles of operation, precision, and accuracy are discussed in detail as well as their power requirements and associated electronics. Observing platforms bearing these instruments cover a diverse spatial range from satellites in orbit, to surface vessels or buoys afloat, to submerged vehicles, and to subsurface and ocean bottom emplacements. Autonomous profiling buoys are now capable of characterizing water column properties from the surface to great depths. Shore-based platforms also provide meteorological data and the novel capability of HF radar surface current mapping for coastal ocean observing. Active and passive electromagnetic sensing instruments aboard satellites in terrestrial orbit provide wide ranging synoptic views of ocean surface and subsurface features. Observing platforms ranging from the traditional to the most recently developed are described as are the challenges of integrating instrument suits to individual platforms.

Operating and maintaining a coastal ocean observing network is subject to the challenges posed to operating electronic instruments and platforms in remote environments where electrical power is unavailable and equipment is subject to harsh conditions. The book describes currently available provisions for reliable power supplies and for protection from seawater pressure, corrosion, and biofouling, provisions which are essential to operational ocean observing.

Large volumes of data are generated by distributed networks of observing platforms constituting an observing system. Depending on the platform, observations from one to several instruments, together with metadata such as time stamps, geolocation, and depth must be integrated into a standardized data packet for transmission to one or more data assembly centers. Electronic data is digitized, filtered, and processed into discrete data packages prior to transmission. Data are then either made available in a few currently accepted data formats or integrated into value-added data visualization products. The book describes the processes involved in data conditioning, quality assurance, and quality control procedures as well as accepted data formats and representative data products.

Data from remote observing sites must be transmitted to the operator. Data telemetry can make use of cables to shore. Data from subsurface emplacements must be transmitted to the sea surface via acoustic means due to the opacity of seawater to radio and microwave frequencies electrical, or via fiber-optic or electrical cable. Telemetry from surface platforms may make use of radio or microwave frequency electromagnetic radiation to shore-based receiving antennae, satellites relays, or, increasingly for coastal platforms, the commercial cellular data networks. Command-and-control of the platform and instruments is likewise effected through such means. Navigation of mobile platforms, and station-keeping assurance of fixed ones, is performed using satellite geolocation for surface platforms or by acoustic means if submerged.

Increasingly, instrumental ocean observations serve to inform continuously running numerical simulations (mathematical models) of state variables such as sea-surface temperature (SST), sea-surface salinity (SSS), sea-surface height (SSH), winds, waves, currents, and ocean color in near real time. Such *data assimilation* schemes constrain model drift extending model prediction skill. Some of the most widely used models and their operational products available through the internet and other applications are discussed.

The process of data dissemination in particular has changed substantially over the last few decades. The majority of data products, including numerical forecasts, can now be retrieved and displayed electronically through access to the internet. Indeed, many of the data products now being disseminated are specifically designed as mobile applications accessible while at sea aboard small vessels over smartphones or handheld tablet computers.

Integrated ocean observing systems operating sophisticated platforms and instruments at sea require significant infrastructure and human resources for sustained systems operation and maintenance and for data archival and product development and dissemination. A final chapter is devoted to the challenges of planning, deploying, and maintaining such systems.

References

Dore JE, Lukas R, Sadler DW, Church MJ, Karl DM. Physical and biogeochemical modulation of ocean acidification in the central North Pacific. PNAS. 2009;106:12235–40.

Edwards CA, Moore AM, Hoteit I, Cornuelle BD. Regional ocean data assimilation. Annu Rev Mar Sci. 2015;7:21–42. https://doi.org/10.1146/annurev-marine-010814-015821. Epub 2014 Aug 6.

Hart JK, Martinez K. Environmental sensor networks: a revolution in the earth system science? Earth Sci Rev. 2006;78:177–91.

Chapter 2
Electronic Sensors and Instruments for Coastal Ocean Observing

Abstract Electromechanical, electro-optical, opto-chemical, and electrochemical sensors are now available that allow continuous real-time monitoring of a wide range of environmental parameters and process rates. These sensors are integrated into electronic instruments capable of directly or remotely capturing these properties or rate processes in quantitative terms as analog or digital data. This chapter describes in detail the wide range of commercially available sensors and instruments with examples for the most commonly measured physical, chemical, and biological variables in the marine environmental field. Principles of operation and limitations of available sensors are also described.

Keywords Transducer · Instrument · Thermistor · Conductivity bridge · Bridge oscillator · Current meter · Current profiler · Anemometer · High frequency radar · Radar tide gauge · Optode · Spectrophotometer · Fluorometer · Wet chemistry · Nutrients

2.1 Transducer-Driven Instruments for Ocean Observing

Transducers are electronic devices that allow measurement of a physical, chemical, or biological property or process. Environmental forcing alters electromagnetic properties of the transducer such as to change its electrical resistance or cause the generation of an electrical potential, mechanical deformation, or electromagnetic emission. Piezoelectric, electromechanical, electrochemical, optical, and acoustic transducers respond to physical, chemical, and biological environmental forcing.

A *passive autonomous measurement instrument* is composed of the primary transducer, an electronic signal amplification unit often incorporating the primary transducer, a computerized instrument control module and data processing unit, a power source, and appropriate signal transmission and reception capabilities. The amplification and processing circuitry provides a readable signal in the form of

---

The original version of this chapter was revised. A correction to this chapter can be found at https://doi.org/10.1007/978-3-319-78352-9_9

J. E. Corredor, *Coastal Ocean Observing*,
https://doi.org/10.1007/978-3-319-78352-9_2

digitally storable readouts. More elaborate *active* measurement instruments require probes to be applied to the target necessitating an acoustic, electrical, or electromagnetic source and associated circuitry for conditioning of the probe signal. Well calibrated, many of these variables can be reported on the scales of the *International System for Weights and Measures* (SI for the French *Système international d'unités*) which govern these fundamental measurements assuring widespread consensus on data accuracy and precision (Bureau International des Poids et Mesures 2006).

An exponential increase in capability of underwater instrumentation has been fueled by the advent of modern electronics. Electronic signal detection and amplification technology was originally developed for radio communications and artillery ranging and detection during the Second World War. Vacuum cathode ray devices that amplify and modulate electronic signals permitted sending and receiving atmospheric radio signals and, subsequently, underwater acoustic signals. Modern devices incorporating *solid state* technology far surpass the performance and reliability of the original vacuum tube and have allowed miniaturization of the components and freedom from the fragile, failure prone vacuum tube technology of 50 years ago. Solid state *transistors*, at the heart of all electronic instruments today, are composed of *semiconductor* mineral phases of materials such as silicon and germanium. *Diodes* (bipolar transistors) consist of a monolithic physical junctions of two such mineral formulations displaying opposite negative (N) or positive (P) electronic properties. Electrical leads to the *source* (positive) and from the *drain* (negative) connect the device to the operating circuit. Diodes permit current flow in only one direction, constituting effective electronic on/off valves that *rectify* oscillatory alternating current to flow in only one direction. Signal amplification transistors known as *bipolar junction transistors* incorporated an additional mineral phase *gate* interposed between the diode elements yielding the configurations PNP or NPN. These *electronic valves*, analogous to the triode vacuum tubes of (recent) yore, allow amplification of the low power signal through *modulation* imparted to a carrier wave. The low power signal energizes the central gate element in a pattern dictated by the sensor and transmitted across the assembly to the drain element both as amplified by the source and as modulated by the gate. Such *power transistors* are recognizable in electronic circuits as those attached to large fluted metal heat sinks. Power transistor heat loss however constitutes a limiting factor for the operation of remote sensors. In practice, these transistors are incorporated into integrated amplification circuits such as the well-known analog operational amplifier. External oscillator circuits feeding op/amps provide frequency modulation. Since the signal from any electronic transducer including acoustic, radio, microwave, and optical emission may be similarly modulated, the application of solid state technology using electronic sensors is extended to many practical ocean observing applications here discussed. In addition to primary data sensing, separate circuitry is required for electronic data conditioning and transmission (Chap. 5).

Today diode- and triode-like logic gate transistors in integrated circuits (IC) with dimensions down to 45 nm can have *transistor counts* of more than 10^9 per IC. The metal oxide semiconductor field effect transistor (MOSFET) and similar designs have proved especially suitable for incorporation into these circuits that are fabricated through photolithographic procedures. In contrast to the original monolithic double junction transistors, incorporating three fused mineral phases (NPN and PNP), a single mineral phase can serve as source and drain. A constriction at the virtual gate is overlain by the *field effect* element where signal charge accumulates or depletes varying resistance across the gate thus modulating the higher power source current.

Paired *complementary* MOSFET units of opposite electronic configuration constitute the so-called *cMOS*, fast logic switches that draw current only during the switching operation minimizing energy consumption. Integrated circuits are now used to condition power for sensor energization, and instrument detection circuitry, to generate and modulate active electromagnetic or acoustic probe signals and to generate and modulate radio frequency or microwave communication signals. Proton-ion selective MOSFET triodes are now integrated into instruments capable of precise, continuous remote pH measurement. *Junction photodiodes*, semiconductor PN junctions sensitive to light, have allowed the design and construction of a wide variety of light-sensing *optoelectronic devices* for optical applications. Silicon photodiodes perform best in the visible region (400–700 nm), while SiC formulations are used for near UV detection (200–400 nm) and InGaAs alloys are used for the near IR band (>700 nm).

The following sections of this chapter are devoted to detailed description of the principles of operation of a wide variety of environmental transducers applicable to ocean observing and to the specific capabilities of various commercially available instruments (referred to occasionally as *sensors*) operating on these principles. Examples of basic transducers as well as circuitry, power requirements and endurance of typical instruments are discussed. Sensors are categorized according to discipline. Instruments measuring physical phenomena including temperate, pressure, winds, waves, tides, and ocean currents are discussed first followed by instruments targeting chemical and biogeochemical variables.

The advanced user is directed to the online publications of the nonprofit US-based *Alliance for Coastal Technologies* (ACT) http://www.act-us.info/ accessed 11/16/2017) for further reference to the subject matter discussed below. The mission of ACT is the evaluation of commercially available sensors for coastal ocean observing. In fulfilling this mission ACT has developed stringent protocols for sensor testing under a wide variety of environmental conditions. Invaluable technical reference is provided by ACT *Workshop Reports, Sensor Evaluations,* and *Technologies Database*.

2.2 Electronic Sensors and Instruments for Ocean Observing

2.2.1 Seawater Temperature

Temperature (T) is perhaps the most widely measured environmental property since, together with pressure, it governs such fundamental properties as the physical state of water and the electrical conductivity of ions in seawater. An intensive property, T depends not only on the heat content of the recipient matter but also its heat capacity which can vary widely. The atmospheric *thermosphere* at altitudes of 100–1000 km with temperatures ranging beyond 2000 °C is in this sense a rather paradoxical example. Its high temperature is due to the absorption of high energy solar radiation by resident gases but the sparsity of these same gases in the rarified atmosphere results in its remarkably low heat content.

Prior to the advent of electronic circuits, the mercury/glass thermometer was the reference for T measurement calibrated by way of two point standards of equilibrated ice/water and steam/water baths. For oceanographic applications where seawater pressure is a concern, elaborate *reversing* thermometers, developed by the instrument manufacturing company of Negretti and Zambra in Victorian England (accurate to 0.01 °C), were deployed in *protected* (glass sheathed) and *unprotected* pairs. These thermometers provided readout by breaking the mercury column in situ. A single loop of the glass capillary terminating in a constriction of the capillary caused separation of the mercury column upon mechanical reversal of the thermometer assembly. The protected thermometer, not subject to local pressure due to the glass sheath, provided in situ T readings. Readout discrepancies between the thermometer pairs caused by *adiabatic* (pressure-induced) warming of the mercury/glass assembly in the *unprotected* unit allowed sample depth estimates independent of the more uncertain estimates provided by the cosine calculation of *wire paid out* and *wire angle* to arrive at true instrument depth (Sverdrup et al. 1942).

Resistance Thermometer Detectors

In practice today, temperature standards are established through the International Practical Temperature Scale 1990 (ITS-90) developed by the SI *Consultative Committee for Thermometry.* Preston-Thomas (1990) prescribes the following standard:

> *The Triple Point of Equilibrium Hydrogen (13.8033 K) to the Freezing Point of Silver (961.78 °C)... In this range T_{90} is defined by means of a* ***platinum resistance thermometer*** *calibrated at specified sets of defining fixed points, and using specified reference and deviation functions for interpolation at intervening temperatures*

The electronic platinum resistance detector (PRTD) calibrated to the standard ITS-90 thus encompasses the full environmental range for liquid ocean water which is roughly −2 to 400 °C (including deep-ocean hydrothermal vents). The PRTD, one of various electronic resistance temperature detectors, was chosen as the SI standard for its long-term stability, largely due to the chemical inertness of the metal (Fig. 2.1). Platinum resistance temperature detectors provide accurate temperature determination for many commercial, industrial, and research applications. For oceanographic use, PRTDs are incorporated into laboratory bench salinometers (see below), CTDs, and flow-through systems.

Thermistor Temperature Detector

Field applications are now shifting to reliance on *thermistor* temperature detectors, temperature-sensitive ceramic or polymeric semiconductor materials that offer faster instrumental response and greater resolution. While the PRTD resistance response to temperature is close to linear, that of most thermistors is steeply curved

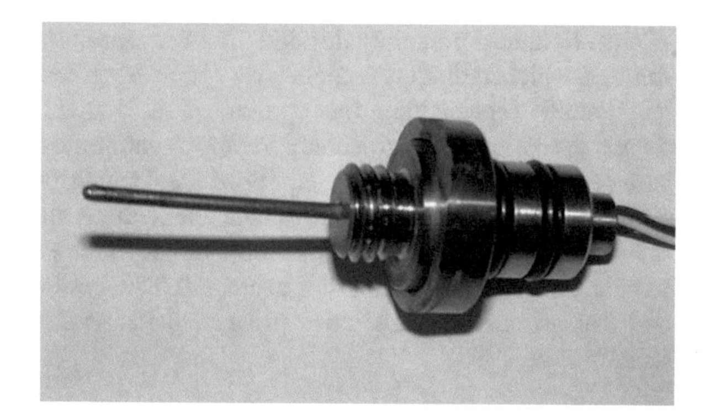

Fig. 2.1 Platinum resistance thermometer integrated in a bulkhead mounting as part of a CTD package

and inverse, with greater resistance at lower temperatures. The pronounced nonlinearity necessitates rigorous calibration and data processing to achieve accurate results throughout the temperature range. Since thermistors are more economical to produce and can provide greater precision and temporal resolution, oceanographic practice trends in this direction although in full recognition that the PRTD remains the primary reference for temperature determination.

Modular electronic temperature sensing units are widely used and are available in multiple application-targeted versions. High precision instruments using aged glass coated thermistor beads calibrated in ITS-90-prescribed baths offer accuracy to 0.001 °C and time responses down to 0.07 s. Electronic means also improve up to tenfold over the accuracy of mercury thermometers. Yearly sensor drift of such instruments can be as low as 0.002 °C. *Wein Bridge oscillators*, employing series and parallel resistor and capacitor pairs (arranged in a diamond pattern reminiscent of the legacy resistance-based *Wheatstone Bridge*) render frequency modulated data allowing intrinsic frequency-keyed digitization of the sensor signal.

Satellite Radiometry

Remote sea-surface temperature measurements from aircraft or orbiting satellites are achieved by means of *pyrometers,* remote sensing instruments designed to measure radiant heat flux taking advantage of *thermal windows* in the infrared and microwave wavelength bands where absorption by atmospheric gases is minimal. These space-borne instruments achieve sea surface pixel resolution on the order of tens (in the best cases) to hundreds or thousands of meters. Microwave channels exhibit poorer resolution than IR bands rendering their products (in the tens of km) of marginal applicability to coastal ocean observing. Infrared instruments aboard NOAA weather satellites have progressed from the *high-resolution radiometer* (HRR) through the *very* high-resolution radiometer (VHRR) to the current third-generation *advanced* VHRR (AVHRR) that afford surface resolution of 1.09 km. Three IR bands (3.55–3.93, 10.30–11.30, and 11.50–12.50 μm) are used for night cloud mapping and sea surface temperature determination.

Spectrometers imaging through *thermal windows* in the infrared (IR) band, through which radiation escapes absorption by greenhouse gases, provide a *skin* temperature representing the uppermost molecular boundary water layer with depths measured in micrometers. Vicarious calibration to in situ temperature measurements taken by thermistor-equipped fixed and drifting data buoys (see below) is employed to derive equivalence to *bulk* temperature, the temperature of the topmost 0.5–2.0 m from which approximate depth buoy measurements are performed. Skin to bulk temperature differentials average 0.3 °K but range between −1.4 and + 1.2 and depend on solar irradiance (time of day), wind conditions, and cloud cover (Emery et al. 1995).

2.2.2 Seawater Pressure

Pressure measurements are performed to estimate depth and to record waves and tides. Initially, commercial electronic pressure sensors employed strain-gauge transducers incorporating a bonded metallic wire grid whose resistance varies with applied stress. Instrumental resolution improved markedly upon the introduction in 1972 of the quartz crystal resonator, a sensor whose frequency of oscillation varies with strain applied to the crystal. Modern CTDs incorporate either type of sensor but the crystal resonator offers higher precision and decreased hysteresis.

Pressure sensors are also used for wave and tide measurements but in neither case are they the primary choice for routine operational observations. Rather they are favored for research and survey in specialized applications such as surf zone deployments.

2.2.3 Ocean Currents

Coastal ocean currents normally span speeds to about 5 knots (2.57 $m.s^{-1}$). However, nearshore currents can achieve significantly higher speeds in narrow inlets and along prominent capes, especially under the influence of large amplitude tides. Tidal bores have been reported at speeds of up to 15 knots. The Bay of Fundy in eastern Canada is an outstanding example. Oceanic mass transport is measured in Sverdrup units (Sv) representing millions of cubic meters per second. The Florida current for example averages 30 Sv, transporting most of the waters exiting the Caribbean Sea and Gulf of Mexico through the relatively narrow Florida Straits between Florida and the Bahama Islands. Surface velocities in certain locations of the Straits can exceed 5 knots.

Historically, two methods have been used for direct current measurement both named after prominent eighteenth-century mathematicians: the *Eulerian* system relying on fixed-point measurements of current strength and direction and the

Lagrangian method in which a parcel of water is tracked by means of a drifting body. Each method has its strengths and weaknesses so that use of either method is governed by the needs of a particular application. For coastal operations, concern is focused on velocity (incorporating a directional component) and dispersion, characterized by stochastic drift of individual Lagrangian bodies embedded in turbulent flow. Stringent operational requirements of ocean observing and the difficulty of deploying and recovering individual drifting bodies favor the *Eulerian* method. A variety of current meters based on different principles are available incorporating electropotential, acoustic, and *Bragg* radio scattering methods. *Lagrangian* methods are not particularly favored for routine operational observation but find use during field campaigns to provide validation for shore-based radio frequency observation of ocean currents to be discussed below.

Eulerian Current Observation

Historic: Ekman, Savonius Rotors In situ *Eulerian* current meters have been a fixture in the ocean sciences for over 100 years since the advent of the *Ekman* current meter in 1903. This ingenious apparatus consisted of a small propeller-like rotor mounted inside an annular duct that protects the rotor and at the same time dampens the effect of vertical movements caused by wave action or platform heave. Propeller rotation was fed to a clockwork mechanism ending in three mechanical dials that recorded units, tens and hundreds of revolutions. A large vane and a swivel mount allowed orientation of the apparatus to the current axis. To measure direction, steel balls were periodically dropped into a cup separated into pie-shaped chambers.

The first modern electronic instrument was introduced in the 1960s by Ivar *Aanderaa* incorporating a *Savonius* rotor, which more efficiently dampened the effect of vertical wave-induced motion, a strong pressure housing, and an electronic package incorporating an *analog to digital converter* (ADC). Digital data received from the ADC was recorded on tape using onboard recorders within the pressure housing. This development can be recognized as a significant milestone in the adoption of digital electronic data systems for ocean observing with greatly increased capacity for data handling and data processing. Despite the great advance provided by the Savonius-rotor meters, on the other hand, these mechanical systems were still subject to corrosion, material wear, and biological fouling and were rapidly replaced by newer technology using systems free of moving parts.

Ocean Current Measurements Through Electromagnetic Induction Physicochemical effects such as those used to measure salinity can be effectively inverted to measure currents. The moving ions in seawater induce current in a stationary orthogonal conductor. Existing undersea cables have been used successfully to document large volume transport. The long-term measurements of Florida Current variability using retired submarine telephone cables (Baringer and Larsen 2001; Peng et al. 2009) are a remarkable example.

On a smaller scale, geometric arrangements of electrodes or induction coils in field instruments achieve similar results. Single-point electromagnetic field induction instruments for field deployment respond to the movement of ions in solution swept along in the ocean current. A durable solution has been found using orthogonal titanium electrode pairs embedded on the surface of a spherical pressure housing to sense horizontal ocean currents. Such instruments reliably measure currents ranging to 750 cm.s^{-1} (about 50 knots) and can achieve resolution thresholds below 0.05 cm.s^{-1}.

Acoustic Current Meters and Profilers Acoustic instruments make use of the piezoelectric effect undergone by materials such as ceramics and quartz crystals that generate an electrical potential upon mechanical stimulus or, vice versa, suffer mechanical deformation upon electrical stimulus. Electrically triggered pulsed deformation of a transducer immersed in an aqueous medium produces an acoustic wave train. In practice, such mechanical transducers are the basis for acoustic communications but have also provided the basis for a wide range of acoustic Doppler instruments. *Piezoceramic* transducers which may be fabricated in various dimensions and designs are widely used for acoustic applications.

Single-point *acoustic current meters* sample small volumes of seawater using high frequency (up to 6 MHz) transducer arrays. Several models in field use incorporate orthogonal pairs or other transducer arrays to provide detailed, extended 3D data time series of current speed and direction. Rapid alternating operation of a transducer pair to either receive or transmit an acoustic pulse allows determination of the velocity vector component parallel to the pair axis by measuring the acoustic phase shift caused by along-axis water transport. Additional orthogonal pairs allow 2D and 3D resolution of the current field. Such instruments can achieve threshold measurements down to less than 0.05 cm.s^{-1} sampling at rates up to 10 Hz. Operating potentials are in the range of 5–32 V DC with typical current drain below 50 mA. Acoustic Doppler *velocimeters* focus four angled orthogonal transducers sampling a volume of 6 mm diameter at 5 cm from the transducer plane (Fig. 2.2). Resolution of such instruments is on the order of 1 mm. s^{-1}.

As described above for single depth acoustic current *velocimeters*, acoustic Doppler current *profilers* (ADCP) rely on return echoes from suspended matter throughout the water column. These ADCPs have found the widest acceptance for extended operational deployment since a single stationary downward- or upward-

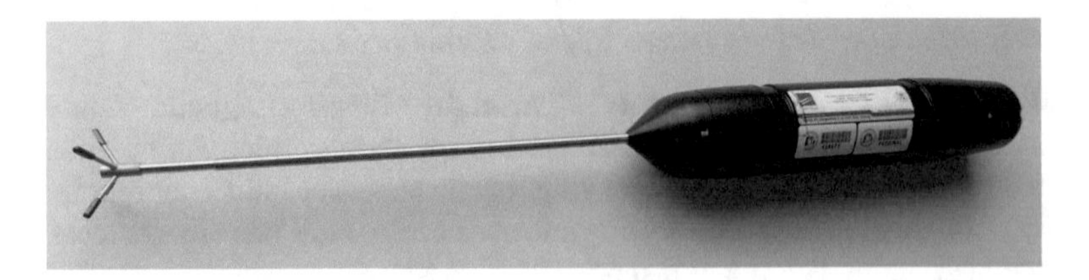

Fig. 2.2 Acoustic velocimeter for single-point current measurement. The instrument measures current velocity in a small volume of water at the focus point forward of the transducer array

looking instrument can provide vertical profiles of current speed and direction to depths as great as 1 km. Instrument designs typically mount three or preferably four transducers in a cloverleaf pattern at inclination of 20°–30° from the axis of the cylindrical instrument pressure housing. Operational deployments typically sample at 0.5–1 h intervals for months to years without need for servicing.

Sound attenuation in the ocean is a function of frequency; as the frequency increases, so does the attenuation. Thus, attenuation of a 10 kHz beam is on the order of 0.5 dB.km^{-1} whereas that for a 100 kHz beam is around 13 dB.km^{-1}, about 26 times greater (Marsh 1969). Acoustic profiling instruments operating across path lengths of tens to hundreds of kilometers mostly operate at lower frequencies (in the range of 10s to 100s of kHz) in contrast to the acoustic current meters discussed above where the path length is measured in centimeters. In operation, each transducer creates a brief coded sound pulse emitting a conical beam and then suspends transmission to capture echoes returning off particles drifting with the current. Repeat-sequence coding using a seven bit code allows increased time/distance resolution (Pinkel and Smith 1992). Sound frequency is down-shifted by particles receding from the instrument or up-shifted by those approaching. Knowledge of the speed of sound through seawater (1400–1530 m.s^{-1} depending on the temperature and pressure) allows calculation of the target distance. Particles reflecting acoustic signals at around 500 kHz are probably zoo- and macro-plankton. Vertical structure and migration of zooplankton aggregations can be effectively characterized using ADCP, but the scatter around the linear response of mixed populations varies to such an extent as to make meaningful biomass estimates difficult (Fielding 2004).

In practice, Doppler data is typically aggregated into depth bins encompassing spans between 0.25 and 20 meters. High power instruments (about 1.5 kW) operating at low frequencies (55/75 kHz dual frequency) can now provide reliable profiles spanning 1 km depth. Hull-mounted instruments aboard ocean going vessels operating at 75 kHz routinely achieve current velocity profiles to depths down to 700 m while underway. Mid-range instruments for coastal applications operating at 300–400 kHz achieve depth ranges of 50–100 m, while high frequency instruments (1–2 MHz) provide fine detail with bin sizes down to 0.1 m over ranges of 10–20 m. A single instrument may be equipped with two sets of transducers operating at different frequencies to take advantage of the fine resolution afforded by high frequencies and the long range achieved with low frequencies. Deep profiles may be achieved on station by lowering the ADCP through the water column and correcting for vessel drift using GPS. For such operations, the ADCP may be mounted in the carousel cage so that the hydrographic profile, water sampling, and the acoustic Doppler profile may all be taken concurrently.

Doppler profilers can be mounted in several configurations including downward-looking such as on buoys and the ship hull mounts discussed above, upward-looking sea bottom emplacements, or sideward looking on pilings or other fixed structures. A combination of the two latter options is optimal for busy harbor operations where swift currents in confined spaces may constitute a hazard concern.

The first one or two data bins in the proximity of ADCP instrument are commonly *blanked out* during the brief transition from the transmit mode to the receive

Fig. 2.3 Acoustic Doppler current profiler. One of three profiling transducers is visible in the head well. Two transducers of the three in the single depth array are visible mounted flush to the housing cylinder

mode. To address this shortcoming, specialized *single depth* transducers emit three or four acoustic beams horizontally from the acoustic head providing data on current velocity at the depth of the sensor thus filling the bin blanked in the profiler mode (Fig. 2.3).

Remote High Frequency Radar Surface Current Sensing The report by Barrick and collaborators in 1977 entitled "Ocean Surface Currents Mapped by Radar," based on a body of observations dating back to 1955, constituted a breakthrough in the field of ocean current measurement allowing the creation of *synoptic* maps of nearshore surface currents *from ground-based stations*. These new *radars* operating in the high frequency (HF) and very high frequency (VHF) radio bands (well below conventional *microwave radar bands*) are known as *high frequency radar (HFR)*. Commercial HFR in the USA builds upon the direction-finding design of Barrick et al. (1977).

The method relies on the phenomenon of *Bragg* scattering whereby radiation backscattered off recurrent features is amplified through positive interference when the radiation wavelength is exactly twice that of the recurrent target pattern and the backscatter angle is exactly 180°. *Bragg* scattering is well known for its application in X-ray crystallography where distances to be measured are on the order of 10^{-10} m. Ocean surface waves exhibit wavelengths on the order of tens of meters, for which distances HF/VHF radio at frequencies from 3 to 300 MHz span the environmentally relevant range. Return echoes, when plotted as absolute power returned (in dB) against displacement from the transmission (Tx) frequency (in MHz), are visualized as *Bragg peaks*, displaced by a few kHz to either side of the central transmit Tx frequency denoting *Doppler* displacement by approaching or receding ocean surface waves. Underlying currents however cause further displacement since the waves are traveling along a moving medium. Since the speed of *deep water ocean waves* is precisely known, the residual *Doppler* displacement can be attributed to the underlying current. In practice, a Tx/Rx pair is emplaced as close as possible to the seashore avoiding interfering structures or vegetation (Fig. 2.4).

Fig. 2.4 Installing a high frequency surface current radar antenna emplacement on the west coast of Puerto Rico. Left: The receiving antenna mounted at the end of a dock. The housing at the top of the support mast holds orthogonal induction coils (see Fig. 2.5). Right: Transmit antenna mounted on the beach. Workers in the background are burying the Rx antenna cable run. Omnidirectional Tx and Rx antenna are mounted vertically and ground planes extend laterally

In newer mid- and short-range systems, a single Tx/Rx antenna is now used but paired Tx/Rx antennae are still required for long-range systems. Long-wave, low frequency HFR operating at 5 MHz is naturally tuned to 30 m wave length ocean waves whereas an intermediate HF range of 12 MHz reflects off 12.5 m waves. Nevertheless, as for many systems here discussed, there is a tradeoff between range and resolution with shorter wavelengths providing greater resolution at the cost of shorter range and vice versa. A horizontal range of up to 200 km from the coast at resolution of 6 km is achieved with the low frequency systems, while high frequency systems can provide resolution down to 0.5 km at ranges limited to 15–30 km. Directional discrimination of return signals received at a single Rx antenna is achieved by means of orthogonal induction coils in addition to the omnidirectional vertical whip antenna (Fig. 2.5).

Only wave fronts directly approaching toward, or receding from, the antenna will produce significant *Bragg* scatter since those traveling at an angle to or parallel to the antennae emplacement will scatter Bragg waves away from the Rx antenna or not at all. Consequently, the current vector resulting from backscatter of radiation emitted by a collocated Tx antenna at any given point within the field of the Rx antenna will be radial to this location (Fig. 2.6).

A Tx/Rx emplacement at a judicious distance (tens to hundreds of km) from the first allows the construction of a second radial plot. Resultant surface current vectors can be computed from the superposition of radial vector fields from the two sites. Distance from the target to the antennae emplacement is estimated by the time delay of the return signal. Resulting data is averaged within bins that can vary from

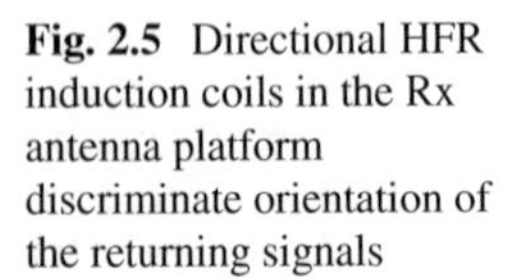

Fig. 2.5 Directional HFR induction coils in the Rx antenna platform discriminate orientation of the returning signals

0.5 km for high frequency/short range systems to up to 6 km for mid- and long range systems. Units operating in the mid-frequency at 13 MHz provide detailed surface short range current maps at 2 km resolution and longer range views at 6 km resolution (Fig. 2.7).

Since seawater is electrically conductive, HF radiation propagates as a *ground wave* that can be viewed as a *half-wave* propagating along the surface and extending at half-wave intervals above the water surface. This property allows transmitted radiation to project beyond the horizon along the geoid and backscattered *Bragg* signals to be detected at the source. Transmit and receive antennae may be fitted with horizontal *ground plane antennae* orthogonal to the vertical omnidirectional antenna to enhance signal transition from the ocean surface to the Rx antenna.

High frequency radar efficacy is highly dependent on salinity since ground wave propagation depends on the electrical conductivity of the aqueous ionic medium. Experimental range determination operating a 42 MHz system in San Francisco Bay was performed by plotting the hourly rate of effective radial returns against salinity. The resulting response is of a rectangular hyperbolic form ranging from a null intercept at practical salinity 3 to a maximum of about 1000 possible returns per hour at salinity above 25. Experiments with a 4.5 MHz *long-range unit* over Lake Michigan, USA, yielded a maximum effective range out to about 20 km, while the range of such a unit operating over ocean water is greater than 200 km. See: https://www.ioos.noaa.gov/wp-content/uploads/2015/12/use_hfr_freshwater_paper2012.pdf (Accessed Sept 9 2017).

An alternate means of achieving directionality using HF is that of the *beam forming* phased-array system. This system offers superior performance to that of the direction-finding design in nearshore applications (Gurgel et al. 1999) but requires

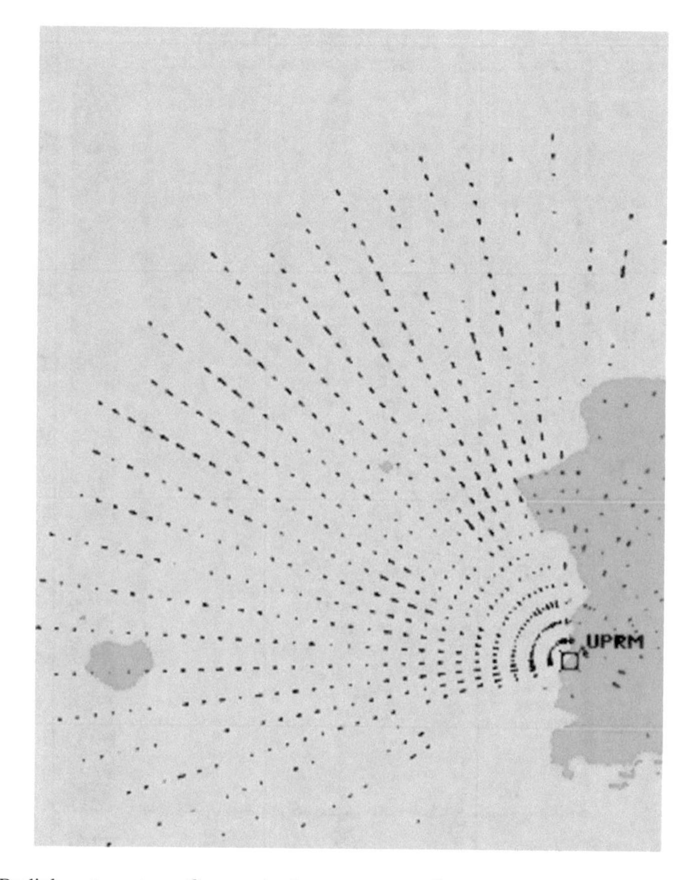

Fig. 2.6 Radial vector returns from a single antennae emplacement

antennae arrays of up to 300 m along the coastline posing aesthetic and regulatory obstacles. These remote ocean sensing systems operate at lower power, typically below 30 W per antenna, and can consist of arrays of up to 12 individual antennae. In addition to ocean surface current mapping, HF radar can be used to locate and track vessels at sea, to make wind and wave measurements, and, under the proper conditions, to detect oncoming tsunami waves (Paduan and Washburn 2013).

Lagrangian Current Observation Lagrangian observations are of moderate use for coastal applications since the spatial scope of water bodies within convoluted coastlines restricts effective coverage to short distances, tens or hundreds of meters in the very nearshore to mid-distances measured in tens of kilometers. Drift of a Lagrangian buoy outside of such a radius of interest can occur within hours to weeks making further data of decreasing usefulness to the question at hand and requiring buoy retrieval and redeployment unless the buoy is purposely expendable. Nevertheless, Lagrangian drifter measurements, coupled to Eulerian measurements and tidal predictions, find useful application for validation of high frequency radar coastal surface current mapping (e.g., Corredor et al. 2011). Dedicated surf zone

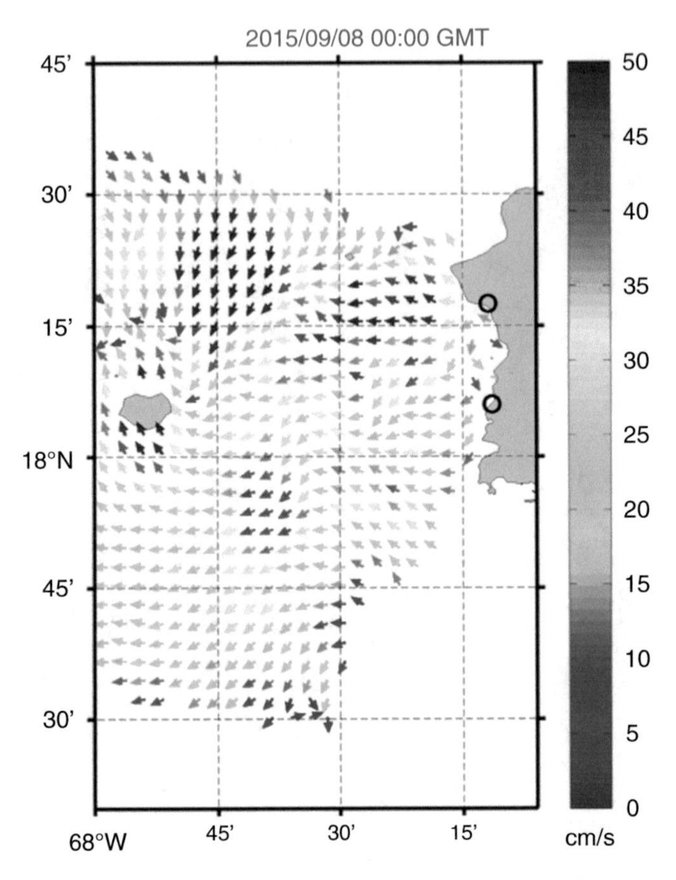

Fig. 2.7 Surface currents in the Mona Passage to the west of Puerto Rico at 6 km resolution obtained from the two HFR emplacements marked by black circles

drifters have been described (Schmidt et al. 2003) but their use is more experimental than operational. Lagrangian methods are further described in the discussion of sensor platforms (Chap. 3).

2.2.4 Ocean Tides

Astronomical ocean tides powered by gravitational forces between earth, moon, and sun and allowed by the fluid nature of the ocean cause the periodic rise and fall of seawater level observed throughout the world ocean except at *amphidromic* points around which the tides are perceived to rotate. Tides are in effect, the *longest waves in the ocean* (Sverdrup et al. 1942). Tidal periodicity varies according to the nature of the astronomical forcing such that a number of partial tides are recognized ranging from the quasi-semidiurnal lunar and solar components through the fortnightly and monthly lunar tides to the solar semiannual component. Since the relative orbital parameters are known with great precision, astronomical tide forecasts are highly reliable.

Nevertheless, poorly predictable atmospheric and geophysical phenomena complicate the forecast. The well-known *acqua alta* phenomenon of Venice Lagoon, discussed elsewhere in this book, is an example of atmospheric forcing which is poorly constrained for numerical forecasting since these atmospheric phenomena are predictable only in the very short term and only with a large body of in situ observations. A well-instrumented observing system established off the coast of Venice, the CNR-ISMAR (Italy's National Research Council Institute of Marine Science) oceanographic Acqua Alta sentinel platform, has allowed some measure of prediction of these harmful events and is undergoing refurbishing at this writing.

Internal waves approaching from deep water break upon encountering a shelf edge and induce a recurrent sloshing of the entire water mass across the shelf, a phenomenon known as *seiching*. Seiching periods are typically on the order of hours depending on the dimensions of the basin which establish the resonance frequency. Seiching is readily apparent in tide gage records as higher frequency sinusoidal oscillations superimposed on the diurnal or semidiurnal astronomical tidal heights. Conversely, internal waves are forced by hydraulic tidal jumps across abrupt ocean bottom sills. Tsunamis, caused by submarine earthquakes, fault displacements, or large-scale seafloor slumping, were once called *tidal waves* but are now recognized as very large amplitude waves unrelated to periodic tides.

As mentioned previously, tide measurement and tidal prediction constitute the precursors of today's ocean observing systems. Tide stations are carefully established with reference to the elevation of an adjoining *benchmark* as a part of a geodetic reference network. Below we discuss the nature and operation of the most commonly used instrumentation.

Staffs and Floaters

Tide staffs are simple ruler-like gauges mounted vertically across the air-sea interface for a direct but crude visual readout of tidal elevation. Tide staffs were normally confined to quiet inshore waters since any wave action would hamper the visual interpretation of tidal height. Tethered floats within a *stilling well* (see Fig. 3.1) attached to a mechanical readout were for many years the most common tide gauges (Sverdrup et al. 1942).

Bubbler Gauges

In the USA, NOOA is responsible for the installation, maintenance, and operation of a National Water Level Observation Network (NWLON). Until the early 1990s, pneumatic *bubbler* gauges were the instrument of choice relying on establishing a stream of bubbles from a compressed gas source emanating from an underwater orifice mid-point along a cylindrical housing pressure open at the bottom. Early operational models (the *Ott* gauge) achieved pressure equalization between the orifice at depth and a surface reference measure (calibrated to the benchmark) with a steel balance beam and a moveable weight operated by an electrical motor. The balance beam served to toggle the motor on and off. Current models incorporate piezoelectric sensors that determine the pneumatic pressure difference between the submerged portion and reference points (NOAA 1991). Although some models used compressed gases in high pressure tanks, current models use ambient air which is filtered, dried with a desiccant, and fed to a pressure vessel by means of an electric piston compressor. These models have the advantage of operating at low current draw using lead-acid batteries and can operate to depths of 35 m (Fig. 2.8).

Acoustic Gauges

Primary tidal observations in US waters are transitioning to acoustic time-of-flight water level gauges that aim an acoustic beam at the air/sea interface within a stilling well and calculate distance from the delay of the return echo. While such

Fig. 2.8 Pneumatic system for bubbler tide gauge. Piston compressor on right, pressure vessel on left, valve body at bottom

arrangements are robust in regions of small tidal amplitude and temperature gradients, very long stilling wells in areas of extreme tides such as the east coast of North America, and regions where near-surface atmospheric temperature gradients can be significant, report large errors.

Microwave Radar Water Level Sensors

The shortcomings discussed above, added to the need for significant infrastructure and maintenance for the acoustic tide gauges, have recently driven the development of free-standing microwave radar water level sensors (Park et al. 2014). These instruments require little more infrastructure than a vertical support periodically referenced to a geodetic benchmark network and dispense with the stilling well although they are still mounted in relatively protected waters where wave action is moderate. The instrument, mounted at a height of a few meters, emits a conical radar pulse at 26 GHz and calculates time of flight of the return pulse. Resulting seawater level is calculated from multiple readings averaging out the wave-induced oscillations.

Implementation of a global ocean observing system, coordinated by the World Meteorological Organization and the Intergovernmental Oceanographic Commission of UNESCO (IOC), has been spearheaded by the development of a Global Sea Level Observing System (GLOSS). In general, technological improvements have mirrored those followed by the NOAA NWLON, but many regions, lagging far behind, still operate float gauges with analog graph paper readout.

2.2.5 *Ocean Waves*

Surface Waves

The ocean surface wave spectrum spans amplitudes of millimeters to tens of meters. A wave field measured by a research vessel in Rockall Trough west of Scotland (Holliday et al. 2006) exhibited a significant wave height of 18.5 m and one individual *rogue wave* was measured at a height of 29.1 m. Wave properties of interest are the height or amplitude of the waves, the direction of propagation, and the wave period, the latter being the time recorded between passages of successive wave crests. Real-world wave fields, however, are chaotic with superimposed wave trains of varying height, direction, and period.

Wave Observation from Surface Platforms

Waves are today most commonly measured using gyroscopically stabilized platforms or chip-mounted microelectromechanical accelerometer devices measuring heave, pitch, and roll together with associated 3D fluxgate compasses for the directional component. Such devices can be mounted on a buoy or a manned or autonomous surface vessel. Buoys designed expressly or primarily for wave measurements are discussed in the chapter referring to observing platforms. Alternatively, stand-alone devices can be custom mounted aboard general purpose ODAS buoys. Buoy-mounted sensors can achieve wave height precision down to about 1 cm but, for operational purposes, wave heights are commonly reported at 0.5 m intervals with the smallest bin thus being 0.5 m (50 cm) amplitude.

Alternatively, buoy-mounted wave measurements can now be achieved using GPS. Remarkably, directional GPS buoys now achieve a precision of 1–2 cm based solely on GPS positioning data. These buoys can be operated in the moored mode but can also be towed or be operated in free-floating Lagrangian mode.

Bottom-Mounted Wave Sensors

Bottom-mounted acoustic instruments also effectively provide wave measurements in the coastal zone. Specialized bottom-mounted ADCPs equipped with dedicated upward-looking wave transducers are available commercially. The upward-looking transducer tracks the air-sea interface while the standard slanted transducers sample the wave-induced orbital motion of near-surface suspended particles. Such instruments, operating at 400 kHz, can be effectively installed at depths down to 100 m. Higher frequencies may be used for shallower emplacements. Resolution is on the order of 1 cm. Manufacturers provide proprietary software to derive the common variables of significant wave height, period, and direction from the acoustic data (Fig. 2.9).

Bottom-mounted pressure sensors were once the preferred solution for wave measurements (Williams 1973), but their use today is restricted to special applications such as the determination of tidal elevation or characterization of surf zone waves where the newer technologies may de inapplicable. These instruments are commonly fitted with high-precision piezoresistive pressure sensors which provide greater resolution than the formerly used strain gauge transducers.

One exception to the above is the NOAA DART tsunami detection system. In this system, an ocean bottom emplacement housing a bottom pressure recorder and an acoustic modem is deployed at basin depths between 4 and 5 km. The modem transmits data at 15–18 kHz to a moored surface buoy with a satellite link. Operational DART systems are deployed along the northwestern Atlantic coast, in the Gulf of Mexico and Caribbean Sea, and in the Pacific and Indian Oceans.

Internal Waves

Internal waves propagate along density discontinuities. The main density discontinuity in the world oceans is the thermocline separating the warm *mixed layer* waters from water masses at greater depth. A pronounced density discontinuity is apparent

Fig. 2.9 Acoustic wave and current Doppler profiler. The central transducer follows the free surface yielding wave height while the lateral array tracks wave-induced orbital motion yielding wave direction. Additionally, the lateral array provides current velocity readings

at the base of the mixed layer at depths of 50–300 m in stratified tropical and subtropical waters. In coastal waters, moreover, buoyant freshwater lenses resulting from river plume dispersal produce one or more haloclines coincident with or additional to one or more thermoclines. Internal wave trains can be induced by hydraulic tidal or current jumps across shallow underwater sills under conditions of exceptional tidal amplitude or current strength and strong internal stratification.

Internal waves propagating along such discontinuities can extend to amplitudes of over 300 m such as those supported by deep thermoclines in open ocean waters. In contrast, waves supported by river plume haloclines in the coastal environment can propagate at depths of less than 30 m with visually apparent rip-slick-rip surface patterns easily detected by common navigation radar. Instrumental observations off Puerto Rico, using ship-based vertical profiling and underwater gliders, document passage of a train of internal waves across the Mona Passage into the Caribbean Sea (Corredor 2008).

Offshore internal wave trains are readily detected by satellite-borne synthetic aperture radar (SAR). These instruments traveling in near-earth slanted polar orbit irradiate a narrow swath of ocean surface with microwave radar beams providing sea- and land-surface relief maps free of cloud interference. The small sea surface displacement caused by the cresting internal wave is readily to apparent SAR. Internal waves can also be detected using stationary manned or mobile autonomous platforms equipped with CTD and or optical sensors.

Shipping interests are usually impervious to the passage of internal waves at sea, but their capacity for inducing seiching oscillations in semi-enclosed basins can affect vessel speed and limit effective ship draft. Nevertheless, routine observations of internal waves for operational applications are not currently implemented, in large part due to the expense of SAR data and shipboard observations.

2.2.6 *Ocean Winds*

Ocean winds, measured up to 320 km.h^{-1} during Pacific typhoons, are largely responsible for surface ocean circulation patterns and their variability. Ocean chemistry and biology are also dependent on these surface winds that influence processes as diverse as gas exchange and larval dispersal. *Anemometers*, instruments used to measure winds, must resist the enormous forces generated by these winds.

Mechanical Rotor Anemometers

In the above context, it is interesting to note that one of the few macro-mechanical sensors used in ocean observing today is indeed the rotor/vane arrangement still used for wind measurements. Streamlined torpedo-shaped, freely pivoting bodies, mounted horizontally on a vertical mast, incorporate forward-mounted helicoidal rotors and vertical tail vanes (Fig. 2.10).

Instruments designed for ocean deployment make use of the most durable corrosion resistance materials such as ceramic bearings for rotor and mast mount. To avoid direct electrical contacts or slip-rings, the rotor carries a set of magnets to energize induction coils within the stationary housing. These current pulses are counted to derive wind speed. For wind direction, rather than using electronic compasses such as used in current meters, these instruments rely on precision potentiometers producing an output voltage directly proportional to the horizontal instrument angle. Prior calibration to a reference compass is required. Such wind gauges are seen aboard many surface platforms such as shore stations, ODAS buoys, and ships. Rotor instruments incorporating conical side mounted cups also are available.

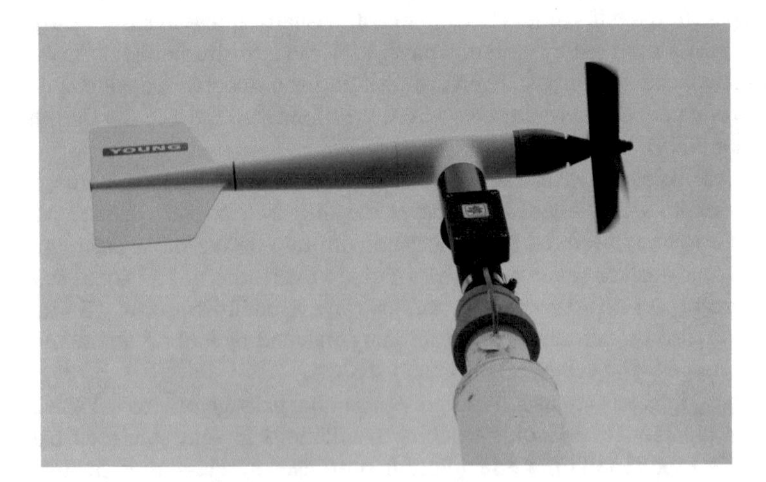

Fig. 2.10 Mechanical anemometer incorporating pivoting body, helicoidal rotor, and vertical tail vane

Fig. 2.11 Acoustic anemometer. Four ultrasonic transducers are visible

Acoustic Anemometers

Acoustic anemometers are similar to acoustic current meters in that they rely on orthogonal paired transducer arrays but dissimilar in that they measure time-of-flight rather than echo return. Taking advantage of their two-way functionality, members of a pair of acoustic transducers alternate their Tx and Rx functions. The time-of-flight difference reveals the net wind component along the array. Readings from the orthogonal pair allow computation of the resulting wind vector. These instruments operate with very low energy draw with a maximum range of wind velocity of 270 km.h^{-1} (Fig. 2.11).

Satellite-Borne Scatterometers

Satellite-borne scatterometers provide an effective means of estimating wind fields over large swaths of the ocean by measuring the return radiation scattered off ocean waves when irradiated obliquely. A patch of ocean is repeatedly irradiated with microwave radiation (C and Ku radar bands) at different azimuth angles, and the *radar cross section* (σ_0), essentially the intensity of the return radiation relative to the irradiation power, is recorded. Absolute intensity provides a measure of the average wave height. The directional component is derived from the differential azimuth readings since, assuming a steady wind field over the course of the readings, for any given wave height, σ_0, will be greatest when the slanted beam points toward incoming waves and minimal when the wave field is orthogonal to the antenna orientation. In practice, a rotating antenna is used to achieve quasi-synoptic differential azimuth readings required for calculation of the vector component. Scatterometer satellites travel in near-earth polar orbits covering swaths of width up to 1800 km with instrument footprints on the order of 25–50 km.

2.2.7 Light in the Sea

Light in the ocean is attenuated through absorption by water molecules and dissolved constituents and through scattering by individual molecules and particulate matter. Absorption is least in the violet/blue region of the spectrum (380–450 nm) and increases at lower and higher wavelengths such that absorption is strong in the ultraviolet (UV) and infrared (IR) bands. Absorption at individual wavelengths is characterized by exponential decay such that

$$I / I_0 = e^{-al} \tag{2.1}$$

where I_0 is the incident radiation, I is the radiation measured after passage of light through a sample over a distance l, e is the base of the natural logarithm (2.718), and a is the absorption coefficient. Across a homogeneous medium, a perfect hyperbola is observed and the absorption coefficient is derived through logarithmic transformation such that

$$\ln\left(I / I_0\right) = -al \tag{2.2}$$

and

$$2.303 \log\left(I / I_0\right) / l = -a \tag{2.3}$$

Light absorption by pure water (Pope and Fry 1997) interferes with the optical detection of dissolved constituents in seawater so that the corresponding absorption of water must be subtracted from the seawater absorption coefficient to obtain the absorption coefficient of these constituents. Since the irradiance ratio is unitless, the resulting coefficient is given as inverse distance. If l is expressed in meters, then the absorption coefficient is expressed as m^{-1}. Normally these measurements are performed upon filtered samples to avoid scattering by particulate matter.

Light attenuation in the ocean medium (c) is given by the sum of absorption (a) and scattering (b). These *inherent* optical properties are mostly measured in the laboratory using spectrophotometers and scatterometers or by means of submersible research instruments available to measure these parameters. Nevertheless, measurement of *apparent* optical properties, dependent on prevailing irradiance, is more suited for practical field applications where the object is to determine the penetration of visible light or photosynthetically active radiation (PAR). *Diffuse attenuation*, incorporating both components, provides a useful measure of light penetration in the ocean allowing the derivation of a diffuse attenuation coefficient denoted as k_d (Kirk 1994).

Photosynthesis is restricted to depths above a *compensation depth* for solar irradiance below which respiration exceeds photosynthesis, a point which is reached at a depth where solar irradiance amounts to about 1% of surface incident radiation. Knowledge of k_d permits computation of this depth. The inverse of the diffuse attenuation coefficient has units of meters (m), a value consequently known as the *optical depth*. This value is of great use for ocean observing since about 90% of the signal received from below the ocean surface by optical satellite sensors emerges from waters above this threshold. Experimental downwelling k_d spectra in the photosynthetically active 412 nm band in eastern Caribbean waters yield optical depths ranging from 0.4 m in the Orinoco River delta to greater than 30 m in oligotrophic waters to the north. More demanding applications of satellite sensors such as the detection of photosynthetic pigments or of dissolved organic matter require instruments capable of quantifying inherent optical properties.

Early measurements of light attenuation in the ocean were performed using the Secchi disk, a white 30 cm disk lowered into the water whose disappearance was visually determined. The advent of *photocells*, incorporating optically active cadmium sulfide photoresistor sensors, rendered the Secchi disk obsolete which themselves were shortly thereafter replaced by the photodiodes that are in use today. Submersible instruments measuring PAR are available in various configurations for incorporation into existing instrument networks or platforms. These instruments are known as *quantum sensors* because their optical range is tailored by means of appropriate photodiodes and glass filters such that there is equal response to photon flux across the PAR spectrum (despite the energy difference of different wavelength photons) and zero response above or below PAR. Light attenuation is commonly measured using paired flat diodes equipped with *cosine collectors* that favor direct incident light from above. Readings from an instrument on or directly below the surface paired to another one at depth allow computation of k_d. Profiling the water column with the lower unit allows more detailed optical characterization of layered water masses (Fig. 2.12).

Fig. 2.12 Cosine collector quantum sensor (PAR meter)

Transmissometers are among the earliest optical instruments available. Equipped with a monochromatic light source that beams to a photoresistor across an enclosed pipe with black walls, the instruments allow single band estimates of k_d. Photodiode-equipped instruments are available today.

Turbidity meters provide semi-quantitative characterization of light scattering. These instruments irradiate a small volume of water with IR and record the incident light resulting from scattering with an orthogonal or adjacent IR-sensitive photodiode. Proxy calibration is achieved with aqueous suspensions of formazin, an organic polymer, and readings are reported as nephelometric units (NTU). Both instruments are useful for estimating visibility and light penetration.

2.3 Electrochemical Sensors for Coastal Ocean Observing

2.3.1 Seawater Salinity

History of Practical Salinometry

Simply stated, the concept of salinity refers to the content of dissolved matter in seawater. Nevertheless, a robust operational definition has been difficult to achieve. Early experiments of boiling the water to dryness were unsatisfactory, attributed subsequently to the chemical complexity of the seawater matrix. It was found for example that carbonate and bicarbonate ions could boil off as carbon dioxide, that organic matter in solution could undergo oxidation as well releasing carbon dioxide, and that absolute dryness was difficult to achieve and maintain due to the hygroscopic nature of many of the resulting salt crystals. For example, after sodium chloride (*halite*), the most abundant salt resulting from evaporation, is in fact magnesium sulfate, the well-known Epson salt, which crystalizes out at *hepta-hydration* resulting in the chemical formula $MgSO_4.7H_2O$ with significant water retention.

In 1901, recognition of the preponderance of the chloride ion in the composition of seawater (55% of total ions in solution by mass), and of the largely conservative proportionality of elements dissolved in seawater (a property known as Marcet's principle), brought about a solution through wet chemical titration of the chloride ion precipitating out as the insoluble silver salt resulting in a derived property termed *chlorinity*. This solution was achieved by means of an international effort advanced by the International Council for the Exploration of the Sea (ICES).

The Practical Salinity Scale and the International Equation of State of Seawater

The *Chlorinity* approach was robust but laborious and was abandoned upon the advent of reliable means of electronic measurement of electrical conductivity resulting then in the *Practical Salinity Scale 1978* under the auspices of the

Intergovernmental Oceanographic Commission of UNESCO and the Scientific Committee for Oceanic Research of the International Union for Science (Lewis 1980). Since electrical conductivity varies widely with temperature, *practical salinity* (S_P) is dependent on accurate measurements of both temperature and electrical conductivity.

Rather than relying on an absolute standard, following the ICES example, the Practical Salinity Scale is based on a *conductivity ratio* between that of the sample and that of an empirical standard supplied commercially to users. This standard known as *Standard Seawater* is prepared using seawater from the North Atlantic Ocean adjusted to a *practical salinity* of 35 (close to the average salinity of the world ocean) against an absolute KCl standard of known conductivity. This conductivity ratio, paired to a concurrent temperature reading of high accuracy, is used to compute practical salinity making use of the empirical equations of the Practical Salinity Scale 1978.

Through careful reevaluation of historical chemical measurement of the individual salts in solution, it is now apparent that the *absolute* salinity (S_A) of standard seawater in mg.kg^{-1} is about 1% higher than the unitless S_P value (Millero 2013). Of late, a new *reference salinity* (S_R) has been defined to bridge the gap in the *Thermodynamic Equation of Seawater* (TEOS-10, 2010) between the unit-less S_P and recognized salt content:

$$S_R = (35.16504 / 35)\,\mathrm{g.kg}^{-1} S_P \quad (2.4)$$

Equation (2.4) can be applied directly to north Atlantic surface water to estimate S_A since this water, by definition, exhibits zero *salinity anomaly* (δS_A). Dissolved silicic acid is in large part responsible for further discrepancy of global ocean salinity derived from conductivity since, at the pH of seawater, it remains well protonated (i.e., ionizes little) and is thus nonconductive. For the different ocean basins, algorithms based on silicic acid content and latitude have been developed for estimating δS_A such that

$$S_A = S_R - \delta S_A \quad (2.5)$$

Large values for δS_A are mostly restricted to deep ocean basins where silicate content is highest. Absolute salinity provides the most accurate estimate of seawater salinity to date (McDougall et al. 2012).

Practical Salinometry

Advances in the instrumental measurement of salinity were made possible by the development of electronic circuitry and, at first at least, by the development of the electrodeless induction conductivity sensor. In essence, this sensor is a voltage transformer consisting of two insulated copper windings and an external conductive core. Alternating current (AC) supplied to the primary winding induces AC in the secondary winding through the pulsing magnetic field generated by the source. In this case, seawater constitutes the transformer core and its ion content modulates the induced voltage in the secondary winding providing a measure of the conductivity ratio to that of standard seawater. This measure, concurrent with a temperature reading, allows calculation of S_P. In early years, laboratory bench salinometers were equipped with such electrodes conformed as two stacked *toroidal* (doughnut-shaped) windings embedded in a nonconductive matrix. Today, some CTDs are also equipped with these toroidal electrodeless induction salinity sensors.

Since the magnetic field of the toroidal cell is external to the sensor, field effects caused by external currents or the proximity of conductive material (such as CTD cages) are significant. Also, biological fouling of the necessarily externally mounted cell causes changes in cell geometry, essential for deriving accurate measurements of conductivity. These drawbacks motivated continuing research to overcome the obstacles to operating immersion electrodes in seawater including electrode drift, external field perturbation, and biological fouling. Design improvements have achieved the necessary resistance stability of the electrodes as well as that of the *cell constant* describing the geometry of the assembly. External field effects have been overcome by minimizing current leak beyond the physical boundaries of the cell. Alternating current is used to prevent changes in seawater chemistry. For long-term deployments, biological fouling is overcome using biocide barriers that prevent the settling and growth of organisms within the small volume cells.

High precision laboratory salinometers allow calibration of field instrument data and analysis of samples where direct immersion of the electrode is impossible (such as sediment porewaters). A widely used bench instrument features four helical platinum electrodes housed in a four-well flow-through glass assembly, all immersed in a constant temperature bath supplemented with a platinum resistance thermometer. Field modules in various configurations are also available. One convenient design incorporates three annular platinum electrodes within a flow-through cylindrical glass cell. Connection of the two outer electrodes conforming a *two-terminal cell* isolates the current field to the interior of the cell and thus prevents current leakage outside the cell boundaries. Submersible instruments incorporate a thermistor thermometer in addition to the conductivity cell for accurate salinity computation (Fig. 2.13).

Remote sensing of surface seawater salinity (SSS) from satellites was only recently achieved through passive radiometric sensing of sea surface microwave emissions in the radar L band at ca. 1.4 GHz. Thermal emission of seawater in this band is modulated by salinity with a change of about 0.5 K per unit change in salinity at 293 °K (20 °C). Negligible absorption of radiation by atmospheric gases in this band creates a satellite remote sensing *window*. Radiometers aboard the European Space Agency *Soil Moisture and Ocean Salinity* (SMOS) satellite (2009–2010) and the US *Aquarius* instrument aboard the Argentinian spacecraft SAC-D (2011–2015) provided proof of concept for eventual operational sensing of the global SSS field (Lagerloef 2012). Nevertheless, spatial resolution of these instruments as deployed in experimental missions is in the range of 100 km^2 largely negating effective coastal ocean observing applications where resolutions of at least 1 km may be required.

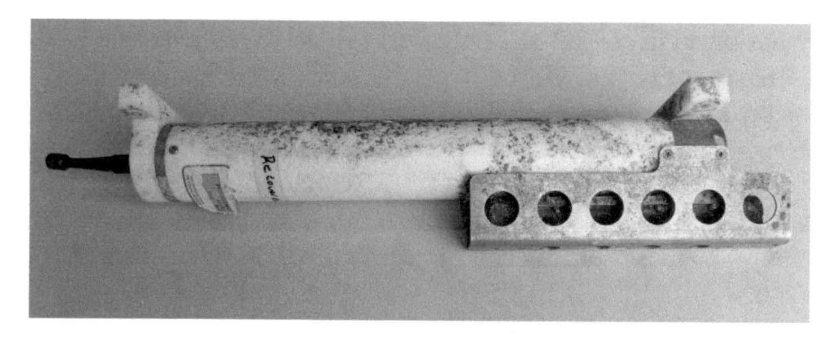

Fig. 2.13 Temperature/conductivity instrument. A cylindrical glass flow-through conductivity cell and a thermistor thermometer mounted externally on the cylindrical housing are protected by the perforated aluminum shield

2.3.2 Dissolved Oxygen in Seawater

Atomic and molecular gases readily dissolve into seawater from the atmosphere where their concentration is a function of their atomic or molecular mass, of their prevailing atmospheric *partial pressure* (the fraction of the total of 1 atmosphere pressure, or 101.3 kPa, exerted by the gas in question) and the temperature and salinity of the water mass. The resulting dissolved gas concentration in seawater is termed *atmospheric equilibrium concentration* (AEC). The bulk of gas in solution is made up of the most abundant atmospheric gases of the atmosphere. Aside from argon (about 1%), heavier gases are scarce in the atmosphere. Oxygen is slightly heavier than nitrogen (32–28 atomic mass units) and is thus slightly more soluble; the proportion of these gases in solution in seawater thus differs from the atmospheric proportion of molecular nitrogen (N_2), about 78%, and molecular oxygen (O_2), about 21%. Solubility is a nonlinear function of temperature, so mixing of water masses with differing temperatures results in supersaturation and spontaneous degassing across the air-sea interface is possible.

Weather-induced changes in local atmospheric pressure and in air bubble dissolution can produce changes in gas concentrations but these tend to be small. By and large, biological and geological effects are most pronounced. For dissolved oxygen (DO), biological photosynthetic production occurs in the surface layers where light is available and slight supersaturation (concentration in excess of AEC) can be observed. Biological consumption on the other hand can occur throughout the water column depending only on the availability of organic matter for food. At geological time scales, production and consumption throughout the world ocean are maintained close to equilibrium by the availability of phosphorus, a key but scarce nutrient and at shorter time scales, by nitrogen availability. Inputs of excess fertilizers and livestock and human wastes have disturbed nutrient cycles in many nearshore environments bringing about excessive consumption of available oxygen resulting in decreased DO content termed *anoxia* (DO absent), *suboxia* (DO <10 μmol.kg^{1}), or *hypoxia* (DO significantly below AEC), conditions that exclude many motile animals and result in metabolic inhibition and decreased scope for growth even for tolerant sessile species. These anthropogenic *oxygen minimum zones* (OMZ) have become widespread in coastal marine waters subject to high nutrient loads brought about by human activity. Enclosed bays and other water bodies lacking adequate circulation are most vulnerable, but even open waters such as those of the northern Gulf of Mexico receiving Mississippi River outflow have for some years now been observed to develop an anthropogenic OMZ. Similar zones are found in other areas of the world ocean. Sustained operational DO measurement to identify OMZ onset and evolution is thus fundamental to coastal ocean monitoring for informed management.

The highest DO concentrations in seawater are observed in polar waters where low temperatures favor gas dissolution, up to 350 μmole.kg^{-1} (11.2 mg.kg^{-1}) at 0 °C and S_P 35 and higher yet at lower salinities. At 25 °C and S_P 35, conditions largely representative of tropical surface waters, DO equilibrium concentration is close to 206 μmole.kg^{-1} (6.6 mg.kg^{-1}). These concentrations can be exceeded by a few percent through the release of oxygen directly into the water as a by-product of photosynthesis but is held in check by degassing to the atmosphere and biological respiration. On the other hand, water masses removed from the surface lose DO through microbial respiration and, in the absence of advective sources, DO concentration slowly decreases. As the *photic zone* (where photosynthesis exceeds respiration) only reaches down to less than 200 m in the clearest waters, the vast majority of ocean waters only receive DO through advective transport of sinking polar waters. Nevertheless, global scale overturning circulation causes ventilation of subsurface water masses overcoming respiration-induced hypoxia such that most of the world ocean is oxic. Oxygen concentrations are lowest in the eastern North Pacific basin where subsurface water masses are oldest. Coastal sediments receive substantially greater loads of particulate organic matter than deep offshore sediments. Bacteria drawing upon oxygen dissolved in the pore waters of sediments bearing high organic loads fuel aerobic metabolism to depletion. As a result, nearshore sediment porewaters are for the most part anoxic. However, burrowing animals can produce channels that increase the oxygen supply to porewaters.

Prior to the advent of modern electronics, oxygen was measured almost exclusively by the *Winkler* titration technique, a complex aqueous procedure involving a redox cascade converting mole for mole, the oxygen fixed in a *pickling* process to molecular iodine, the final oxidant in the cascade which is titrated with a thiosulfate solution of precisely calibrated concentration. The Winkler method is of great precision and is still used today for the calibration of electronic field deployable sensors. In current practice, automatic titrators equipped with optical or electrochemical end-point detectors are used for the final step, but the method remains slow and laborious and is unsuited for continuous flow-through measurement.

Compact field deployable electronic instruments capable of DO measurement within the environmental range based on electrochemical and optical sensors now allow extended autonomous deployment. Modular units are built for integration into multiprobe instruments in wide configurations. *Polarographic* and *optode*-based DO measurement systems are discussed below. Since both types are sensitive to temperature as well, appropriate sensors can be included in the instrument package if not incorporated directly into the sensor unit.

Polarographic Oxygen Electrode

The *polarographic* (amperometric) oxygen electrode or *Clark*-type electrode constituted the first practical electronic means of measuring DO concentration and advanced models remain in use today. The enabling reaction is the electrochemical reduction of DO to hydroxide as follows:

$$O_2^- + 4e^- + 2H_2O \rightarrow 4OH^- \tag{2.6}$$

The theoretical redox potential of natural oxygenated waters versus the standard hydrogen electrode, largely dominated by the oxygen reduction reaction, is a little over 700 mV and varies little despite wide variations in DO concentration. Application of a potential in excess of this value will result in the above reaction and measurement of the current flowing through the cell provides a measure of oxygen concentration. In practice an overpotential of about 800 mV is applied across an electrode pair (usually a platinum or gold cathode and a silver anode). Electrons dispensed to the aqueous medium at the cathode are provided through oxidation of the silver anode with the subsequent deposition of silver chloride (a largely insoluble salt) upon its surface. The reaction cell is confined in a small volume with the electrodes immersed in a potassium chloride (KCl) electrolyte. A gas-permeable membrane separates the cell from the seawater. Cellophane was once widely used as the membrane material; Teflon® is currently preferred. Consumption of DO within the cell creates a gradient and prompts diffusion of DO from the seawater medium to the cell. Hence, the cell current is dictated by the rate of diffusion of DO through the membrane to the cathode site and this in turn is dictated by the outside concentration. The resulting measurement is known as the *diffusion current.* In addition to the raw current reading, calibration requires knowledge of temperature and local pressure.

Initial instruments proved to be unsuitable for high precision oceanographic work requiring that they be *continuously calibrated against Winkler analyses of simultaneously collected samples* (Atwood et al. 1977). This drawback is in large part a result of the continuous change in chemical properties of the electrode known as electrochemical drift and is dependent upon the amount of DO in the sample and the time transpired in operation. Moreover, poorly designed cells suffer significant pressure deformation inducing *hysteresis*, in which the instrument reading becomes a function of the past state of the cell. Thus in vertical instrument casts at sea the *downcast* profile differs significantly from the ensuing *upcast* profile. Membrane fouling is a further cause for concern. Refinements in instrument design, such as reducing cell volume and pressure-induced cell volume changes, reducing membrane thickness to facilitate oxygen diffusion and thus increase response rate, and in operation strategies, such as decreasing operation time to reduce electrochemical drift, have largely minimized (but not eliminated) these drawbacks. Oxygen depletion due to consumption by the cell has been largely overcome by actively pumping seawater across the cell.

Commercial field instruments have ranges to 120% of AEC and initial accuracy of about 2% of saturation or close to 8 μmole.kg^{-1}. Robust stability (0.5%) makes this instrument capable of deployment for over about 4 months, but the manufacturer stresses that a clean membrane is required, a difficult feat to achieve in highly productive coastal environments or near-surface waters subject to oil pollution. Response time varies 8–20 s, making it unsuitable for detecting rapid changes such as fast vertical profiling of natural waters but useful for medium-term mooring applications where burst sampling of a few minutes per hour is common for saving power and reducing data density.

Oxygen Optode

Electro-optical fluorescence-based sensors (known as *optodes*) that bypass many of the difficulties of the amperometric approach have demonstrated improved performance and are well on their way to being the primary method for DO measurement. Fluorescent materials are organic compounds that absorb visible or ultraviolet radiation and subsequently lose the resulting excited state through reemission at longer wavelengths. Initially developed for medical applications (as was the amperometric electrode), the optode makes use of the effect of oxygen on the fluorescence emitted by light-absorbing organic metal complexes providing a novel means of measuring DO without consuming it. Porphyrin complexes of platinum and other transition metals emit red light following excitation (Mills 1997). This emission is modulated by the presence of oxygen which impedes or *quenches* fluorescence. Thus fluorescence is greatest in the absence of DO and least at high DO concentrations. In practice such a complex embedded in a sensing foil target exposed to the aqueous medium is irradiated with blue light provided by a LED device and the resulting red emission is detected by photodiode. Modulation of the excitation signal and measurement of the phase of the emitted radiation allows a measure of fluorescence decay time, further increasing precision. An additional red LED is used as a stability reference source. The sensing foil is provided with a black optical isolation coating to block stray ambient light from reaching the photodiode. Initial response times for current generation field instruments now stand below 30 s with accuracy below 8 μmole.kg^{-1}.

Under the auspices of Alliance for Coastal Technologies, field tests were performed on three commercial optode instruments and one pulsed polarographic sensor using Winkler titrations for reference under widely varying environmental conditions at seven sites in US waters varying from coral reefs to the Great Lakes to inshore waters of Chesapeake Bay (see ACT evaluation reports). Stability during deployments of 4 weeks, sampling at 5 min intervals, was excellent for all instruments in some low productivity environments but most instruments failed catastrophically, within a week in one case and within two in others, when deployed in more challenging environments. *Biofouling prevention systems* (see Chap. 4) fitted to some instruments by their manufacturers showed mixed results as well. Some covaried with unprotected instruments of the same model throughout the deployment while others in more demanding environments allowed extended operation for days to weeks. Some protected versions failed well before the unprotected version but others showed marked performance improvement.

In general, results were encouraging for month-long deployments in certain low productivity environments such as the coral reef waters of Kaneohe Bay, Hawaii, or those of the test area in the US Great Lakes. On the other hand, deployments in environments with very high biological fouling rates cannot be relied upon for more than a few days. One instrument that otherwise showed acceptable performance reported a drop of 8.6 mg.l^{-1} or 268 μmole.l^{-1} in a high biological fouling environment of the Chesapeake Bay.

Both instrument types are subject to biological fouling, be it of the polarograph membrane or the optode foil. Biological fouling interferes with gas exchange and can even constitute an oxygen sink in sufficient biomass. Oil slicks are not encountered as often but oxygen absorption is similarly impeded when it coats the active element.

Despite the advances related, performance of these instruments leaves space for improvement. Response times of even the faster amperometric instruments are still unsuitable for rapid profiling and neither those nor the currently available optodes can provide accuracy much greater than about 8 μmole.kg^{-1}. Carefully executed Winkler titrations using automatic titrators provide accuracy about four times greater at around 2 μmole.kg^{-1}. Winkler calibration using discrete samples, judiciously timed and spaced, is recommended for applications requiring high accuracy and precision. Coppola et al. (2013) have provided an in-depth review of sensor accuracies and scientific need.

2.3.3 *The Inorganic Carbon System and pH in Seawater*

The pH scale provides a shortcut for expressing the acidity of liquids which can vary over 14 orders of magnitude. This acidity is caused by ionized hydrogen atoms or free protons in solution (H^+) which combine with a water molecule to form the ionic species H_3O^+. Since this concentration can effectively range from as little as 10^{-14} mol.kg^{-1} (strong lye) to as much as 1 mol.kg^{-1} (battery acid), an inverse log scale is used with resulting pH values of 14 and 0. Distilled water is by definition *neutral* (neither acidic nor alkaline) at pH 7, while seawater is slightly alkaline with over ten times less protons in solution.

Anthropogenic release of carbon dioxide (CO_2) into the atmosphere, primarily due to combustion of fossil carbon, is causing *ocean acidification* (Caldeira and Wicket 2003; Zeebe 2012) because CO_2 reacts with seawater upon dissolution to form carbonic acid, which rapidly dissociates to carbonate, bicarbonate, and H^+ ions. Historical data indicate that a net pH decrease from 8.2 to the current 8.1 has occurred since the dawn of the industrial era, a time span of less than 300 years. It is expected that continued fossil C emissions will result in a net decrease of 0.2–0.4 pH by the end of this century (Caldeira 2007) with eventual stabilization at about pH 7.6. Sustained observations at the Hawaii Ocean Time Series (HOT) and other oceanic time series have already demonstrated a consistent drop of 0.04 pH in open ocean surface waters over a period of 20 years (Dore et al. 2009).

The above numbers are modest at first sight due to the logarithmic transformation. Nevertheless, when transformed to absolute numbers, the observed decrease from 8.2 to 8.1 represents an increase of 26% in acidity since the 1800s, while the end-of-the-century forecast (to pH 7.8) represents an increase of 76% and the long-term projection to pH 7.6 represents a tripling of the amount of acid in ocean waters. The latter changes have not been observed in the geological records throughout the last 400,000 years (Sarmiento and Gruber 2006).

Chemical equilibrium is established between the atmospheric and gaseous phases and the subsequent reaction products as follows:

$$CO_{2(g)} = CO_{2(a)} \tag{2.7}$$

where $CO_{2(g)}$ is the gas in the atmosphere and $CO_{2(a)}$, the dissolved aqueous species. The latter in turns reacts with water to form carbonate, bicarbonate, and H^+ ions:

$$CO_{2(a)} + H_2O = H^+ + HCO_3^- \tag{2.8}$$

and

$$HCO^-_3 = H^+ + CO_3^{2-} \tag{2.9}$$

Stoichiometric constants for these reactions in seawater are

$$K_H = pCO_{2(g)} / pCO_{2(a)} \tag{2.10}$$

where p denotes the partial pressure of carbon dioxide,

$$K*_1 = \left[H^+\right] + \left[HCO_3^-\right] / \left[CO_2\right] \tag{2.11}$$

and

$$K*_2 = \left[H^+\right] + \left[CO_3^{2-}\right] / \left[HCO_3^-\right] \tag{2.12}$$

where $K*$ denotes apparent equilibria under the particular conditions of seawater T, S, and P and the brackets denote the concentration in seawater of the chemical species (including dissolved complexes with other ions). Millero et al. (2002) have developed empirical expressions for the derivative log transformed constants $pK*_1$ and $pK*_2$ based on experimental data for the environmental ranges of T, P, and S. At $S = 35$ and $T = 25$ °C and $P = 1$ atm, these values are around 6.0 and 9.1. Since these values bracket the current average pH value of about 8.1, seawater constitutes a powerful acid-base buffer.

Despite the buffering properties of the carbonate system, increased CO_2 injection results in reduced pH and depletion of the carbonate ion at the expense of newly formed bicarbonate ion. Simple calculations with the above expressions show that, at the current ocean carbon content and at an assumed pH 8, the ion ratio HCO_3^-: $CO3_2^=$: $CO_{2(a)}$ is about 100:10:1 whereas the ratio changes to 100:1:10 upon acidification to pH 7. That is to say that upon acidification bicarbonate concentration is proportionally affected little since it is the most abundant species but carbonate decreases dramatically, by a factor of 10 over 1 unit pH change, as carbon dioxide increases.

In the ocean, acidification threatens one of the most fundamental pillars of plant and animal life: the *calcareous* skeleton composed of the mineral salt calcium carbonate ($CaCO_3$). Calcium carbonate in surface seawater is normally supersaturated (see below), but the degree of saturation varies greatly in the ocean. Thus, under certain circumstances discussed below, this calcareous skeleton can be subject to dissolution. The tendency for a sparingly soluble salt such as $CaCO_3$ to dissolve is numerically represented by its *solubility product* K_{SP}:

$$K_{SP} = \left[Ca^{2+}\right]_s \left[CO_3^{2-}\right]_s \tag{2.13}$$

In the above expression, concentrations denoted by the subscript s are those found experimentally at exact saturation of the salt, achieved in the laboratory by adding salt to water until no further salt dissolves and a solid phase remains in equilibrium with the dissolved phase. The degree of saturation of a salt, denoted as Ω, is given by the ratio between the so-called *ion product* (IP, the product of the calcium and carbonate ion concentrations observed in a sample) and the corresponding K_{SP} for the observed T, S, and P. For calcium carbonate, the *saturation index* Ω is given by

$$\Omega = IP / K_{SP} \tag{2.14}$$

where

$$IP = \left[Ca^{2+}\right]_{sw} \left[CO_3^{2-}\right]_{sw} \tag{2.15}$$

where the subscript sw denotes ions in solution in seawater. Thus, if Ω is greater than one, the solid phase is stable but, if Ω is lower than one, the salt will tend to dissolve. In the warm, high salinity surface waters of the central ocean gyres, $CaCO_3$ is well saturated with Ω varying from around 5 to 10 (depending on the crystal form, either calcite or aragonite) whereas, due to pressure effects, cold deep abyssal waters are undersaturated with Ω values below 1.

Total inorganic carbon (C_T) is given by the sum of the concentrations of the three components of the inorganic carbon buffering system, an expression known as the *mass balance equation*:

$$C_T = CO_2 + HCO_3^- + CO_3^{2-} \tag{2.16}$$

C_T can be measured accurately using an instrument known as a Single Operator Multiparameter Metabolic Analyzer (SOMMA). In the world, ocean C_T averages around 2 mmolC.kg^{-1}.

The *charge balance equation*, incorporating the major acid/base ionic species in solution in the ocean, yields the *alkalinity* or acid depletion of seawater:

$$A_T = \left[HCO_3^-\right]_T + 2\left[CO_3^{2-}\right]_T + \left[B(OH)_4^-\right]_T + \left[OH^-\right]_T + 2\left[PO_4^{3-}\right]_T + \left[HPO_4^{2-}\right]_T + \left[SiO(OH)_3^-\right]_T - \left[H^+\right] - \left[HSO_4^-\right] - \left[HF\right] \quad (2.17)$$

Subscript T denotes the sum of ions. Since the proportional composition of the major dissolved components in seawater is largely conservative, knowledge of salinity allows their accurate estimation and, by difference, computation of the carbonate alkalinity:

$$A_C = \left[HCO_3^-\right] + 2\left[CO_3^{2-}\right] \quad (2.18)$$

A_C can be accurately and reliably estimated through total seawater alkalinity titration. At S 35 T 25 and P 1 atm, A_T amounts to 2.48 meq.kg^{-1}. Since one *equivalent* of charge represents one *mole* of charge, carbonate contributes twice as much as bicarbonate on a molar basis. Nevertheless, bicarbonate concentration exceeds that of carbonate by a factor of about 10. For the highest accuracy, current oceanographic practice involves extensive calibration with *reference materials*, seawater samples of known DIC and A_C (Feely et al. 2001).

At least a dozen software computational packages are available to compute the different components of the system as expressed above from two such measurements together with T, S, and P (Orr et al. 2015). Instruments capable of repeated autonomous measurement of pH, $pCO_{2(g)}$, and $pCO_{2(a)}$ now allow reliable extended deployment aboard diverse platforms.

Many marine organisms take advantage of this supersaturation of calcium carbonate in near-surface seawater ($\Omega > 1$) to maintain durable calcareous exoskeletons at little metabolic expense. In extreme cases, as ocean pH decreases, near-surface seawater can become *corrosive* to calcium carbonate as are cold deep waters under high pressure. Deep waters of the northeastern Pacific Ocean are highly acidic due to extended residence below the sunlit photic zone where microbial respiratory activity releases CO_2. Furthermore, anthropogenic CO_2 released into the atmosphere in northeast Asia is thought to be absorbed by sinking ocean waters and transported eastward to upwell along the North American seaboard. Already, increased upwelling of corrosive waters with Ω_{arag} (aragonite saturation index) < 1.0 and pH values <7.75 off the west coast of the USA has been observed (Feely et al. 2008). These changes significantly affect the shellfish industry since late-stage larvae show carryover effects of early exposure to low pH (Barton et al. 2012). In view of its grave effects on the oyster mariculture industry, the Central California Coastal Ocean Observing System (CENCOOS), in collaboration with academia and private industry, operates a digital *ocean conditions dash board* online for commercially important Humboldt Bay with real-time pH data and an upwelling index and forecast. http://www.cencoos.org/data/humboldt/oyster (accessed Sept 15 2017). Similar efforts are underway in Alaska (http://www.aoos.org/alaska-ocean-acidification-network. accessed Sept 15 2017).

Sensitivity of reef corals to ocean acidification has been largely attributed to decreases in the aragonite saturation index rather than directly to the increased acidity (Langdon et al. 2000; Caldeira 2007). Nevertheless, this concept has lately come under increased scrutiny since intracellular fluids are actively controlled by the organism. An alternative *proton flux limitation model* (Cyronak et al. 2015) addresses the problem in terms of environmental limitations upon biochemical proton pumps that favor precipitation by altering carbonate chemistry at active intracellular precipitation sites.

Regardless of the biochemical mechanism, increased acidification is known to be detrimental to a wide range of marine organisms. The adverse reactions of pteropods, a planktonic mollusk, to ocean acidification is of special concern since this organism constitutes a prime prey for various Pacific salmon, species of great commercial importance. Profits from shellfish culture and ecosystem services provided by coral reefs are other examples of the losses that may accrue and detrimental economic effect of ocean acidification. Another, less obvious effect is the *noisier ocean* resulting from changes in boron acid-base chemistry reducing its sound attenuation capacity.

The complexity of environmental pH control by the carbon system requires the measurement of at least two variables for its full characterization that is the ability to estimate the value of all remaining variables. Those that can be measured in the laboratory are pH, $pCO_{2\,(g)}$, $pCO_{2\,(a)}$, total inorganic carbon (C_T – see below), and total alkalinity (A_T). Together with concurrent measurements of S, T, and P, these data pairs provide an estimate of the remaining and derived variables. Different analytical pairs however result in different precision of the various estimates. Autonomous in situ measurement instruments are at time of writing only available for pCO_2 and pH. This constraint significantly reduces the options for autonomous instrumental measurement. Since pCO_2 and pH tend to covary, precision of derived values is poor. Development of autonomous instruments for measurement of C_T and AC will allow greater precision. An overview of the theory, principles, and practice of operation of the instruments available for autonomous measurement of the two components of the inorganic carbon equilibrium system, pH and pCO_2, follows.

pH Sensors for Ocean Observing

The current IUPAC definition of pH is an empirical one and, for historical reasons, is based on the *NBS definition* (after the former US National Bureau of Standards) where it was developed. Measurement is based on determination of the potential electromagnetic force (EMF) developed at an *ion selective electrode* half cell with reference to a known standard half-cell. The glass membrane electrode, more properly a half cell, familiar to most who have taken a basic chemistry course, is the most common electrode sensitive to H^+. A thin glass bulb houses a platinum electrode immersed in a potassium chloride (KCl) solution. A potential proportional to the difference in pH is generated across the glass membrane. Standard half cells (also referred to as electrodes) of known potential are the calomel electrode and the silver/silver chloride electrode. Measurement of seawater pH using this approach has proven to be imprecise and variable, depending on salinity of the sample and the architecture and history of individual cells. A *liquid junction* between the reference half-cell electrolyte and the seawater medium is required to complete the cell circuit, but the high ionic strength of seawater causes the development of a liquid junction potential that is at present both undeterminable and irreproducible. A significant fraction of the protons in solution in seawater exists in the form of ion pairs such as HSO_4^-, a situation vastly different from the low ionic strength NBS standards which discrepancy necessitated the development of a *total pH* scale incorporating the singly protonated sulfate (Dickson 2010). Special high salinity buffers (Dickson 2010) are required to properly calibrate a cell in order to avoid large unknown *liquid junction potentials*. In addition, fouling of the aqueous junction remains problematic although porous Teflon appears to resist biological fouling and chemical precipitation.

Dye-Based pH Spectrophotometry

Poor resolution and poor reproducibility of the potentiometric approach for seawater applications using the glass electrode has driven other approaches to pH measurement favoring a colorimetric dye solution. Clayton and Byrne (1993) described a simple procedure involving the addition of an organic dye (m-cresol purple) to a seawater sample and measurement of the resulting light absorption at three wavelengths of visible light; two variable peaks representing protonated or unprotonated dye and a valley in between. These measurements together with precise T and S data allow precise and reproducible estimation of pH. Automation of the procedure has allowed the production of field deployable instruments. Optimized dyes are combined with seawater and subjected to colorimetric analysis (Martz et al. 2003; Seidel et al. 2008). These, and similar autonomous *wet chemistry* instruments discussed below, incorporate pumps, valves, light sources, and detectors with substantial power demands that can be met with robust battery packages and pulsed operation to extend deployment endurance to over 6 months. Response time is also rather sluggish, up to 3 min, due to the need for sample/dye equilibration and sample cell flushing. Improved accuracy and precision and low instrument drift constitute substantial improvements over the glass electrode (Fig. 2.14).

Solid-State Ion-Selective Field Effect Transistors for pH Measurement

A solid-state solution to electrochemical determination of pH free of the vagaries of the glass electrode had been available for some time now in the form of the *ion selective field effect transistor* (ISFET) developed by Honeywell for the food and

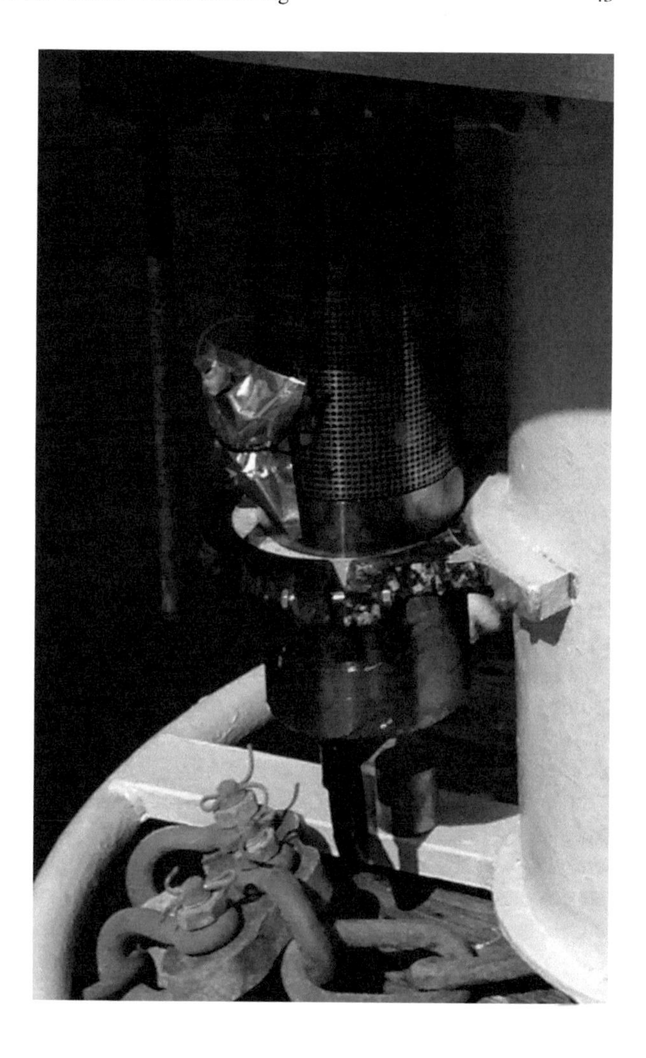

Fig. 2.14 Dye-based field pH meter installed on a data buoy

cosmetic industries. This proprietary technology has now been adapted for autonomous seawater pH measurements through collaboration between academia, a private foundation, and industry (Martz et al. 2010; Johnson et al. 2016). Rugged and durable, with a faster response than glass, in this solid-state device electron flow between source and drain is modulated by the potential difference of a proton-sensitive gate electrode. Uniquely, this electrode is activated by a *counter electrode* that ensures a constant current across the ISFET (Branham et al. 2017). A particularly novel solution to the difficulties caused by the liquid junction between the seawater medium and the reference electrode has been the substitution of the problematic silver/silver chloride electrode with a solid-state ion selective chloride electrode directly exposed to the seawater (Takeshita et al. 2014). Since chloride is the major single component of seawater, a stable, accurate signal is easily calibrated to paired seawater conductivity and temperature data.

Field tests coordinated by the Alliance for Coastal Technologies of five glass electrode instruments, one dye-based sensor, and one ISFET instrument have largely confirmed these advances. The combination ISFET/chloride electrode demonstrated robust performance in the marine environment. The colorimetric instruments and a single glass electrode also showed good precision but greater uncertainty. Glass electrode performance varied widely with only one of the five demonstrating comparable performance to the newer approaches.

pCO_2 Sensors

Automated measurements of CO_2 partial pressure (pCO_2) rely either on spectroscopic determination of infrared radiation (IR) absorption or spectrophotometric dye-based applications. The former use nondispersive IR (NDIR) photometers measuring gas absorption in the 4.3 μm band relative to a reference band at 4 μm. Ceramic-coated resistances generate the radiation that is propagated across a gas cell to an IR-sensitive photodiode. Additional sensor channels to measure water vapor which also absorbs IR may be added to eliminate the necessity for prior drying of the gas stream. Modular gas analyzers are widely available.

For atmospheric measurements of $pCO_{2(g)}$, the gas analyzer cell is flushed by a pump and valve arrangement alternating samples with calibration gases. Chemical dryers may be used to remove water vapor from the sample stream. For continuous determination of surface seawater $pCO_{2(a)}$, an air well may be used for internal equilibration of seawater pCO_2 with a near-surface air stream across a gas permeable membrane. In surface instruments the membrane may be exposed directly to ambient water. Buoy-based field instrument performance evaluations by ACT over an effective range of 300–800 μatm pCO_2 show enhanced performance for those NDIR instruments using onboard calibration with a *zero gas* (achieved by chemically removing all CO_2 from an air sample) and a further *span reference* mixed gas preparation such that the sample values are bracketed within these two standards. Biological fouling was found to be well controlled by the use of copper foil, wire mesh, or perforated sheets protecting the active membrane or intake. Air wells on buoy-mounted instruments are purposely constructed using a nickel copper alloy with anti-fouling properties.

Shipboard systems may use a showerhead dispensing seawater spray into a cylindrical housing flushed with boundary layer air pumped from the ship's bow for equilibration rather than the membrane approach. This air equilibrates with the seawater and is then piped across a drying tube to a NDIR detector. Dry near-surface air piped directly to the detector without prior equilibration serves to determine atmospheric $pCO_{2(g)}$. Shipboard shower head systems and the buoy-mounted membrane system achieve greater accuracy by periodically measuring a blank and a so-called *span* reference sample from an onboard tank dispensing a very precisely measured calibration gas. Since gas concentration in the gaseous state varies with water content and atmospheric pressure, systems measuring the gaseous phase report concentration as the mole fraction (xCO_2). The sample is assumed to be saturated with water vapor which mole fraction can be computed from *T*. Partial pressure, (pCO_2) is then computed with knowledge of atmospheric pressure.

Alternatively, dye-based methods rely on equilibrating CO_2 between seawater and a bromothymol blue dye solution across a gas permeable membrane, thus changing the pH of the dye solution, and subsequent measurement of its resulting optical absorption. Color changes are monitored at the diagnostic peaks of 440 and 620 nm referred to a nonabsorbing blank at 715 nm (DeGrandpre et al. 1995). Current system design allows deployment for up to 1 year autonomous operation with sampling at hourly intervals. Concurrent deployment of automated pH measuring systems such as those described above allows explicit solution of the carbonate equilibrium equations discussed above (Grey et al. 2012).

Carbon Dioxide Measurements from Space

Placement in orbit of three satellites capable of measuring atmospheric CO_2 from space, the Japanese Greenhouse Gas Observing Satellite (GOSAT), launched in 2009, (https://www.env.go.jp/en/focus/docs/files/20120201-26.pdf accessed 11/13/2017); the US Orbiting Carbon Observatory-2 (OCO-2), launched in 2014, (Eldering et al. 2017); and the Chinese Carbon Dioxide Observation Satellite (TanSat) launched in 2016 (Liu et al. 2013), is now affording synoptic views of spatial atmospheric CO_2 distribution with unprecedented resolution. In general, spectrometers aboard these satellites measure sunlight reflection by atmospheric CO_2 in narrow near and mid-IR CO_2 spectral absorption bands. While these satellites do not measure CO_2 in seawater, concurrent in situ measurements can help refine estimates of sea-air CO_2 exchange. Indeed, Chatterjee et al. (2017) have used OCO-2 data in conjunction with seawater CO_2 data from NOAA ocean buoys to *unravel the timing of the response of the ocean and the terrestrial carbon cycle during the 2015–2016 El Niño*. Regionally pertinent data is afforded by the OCO-2 with a footprint of nominally 1.25 by 2.4 km.

2.3.4 Inorganic Nutrients Dissolved in Seawater

The lovely blue transparent waters of tropical beach resorts owe their pristine look to their dearth of dissolved inorganic nutrients (plant fertilizer) that makes them virtual ocean deserts. Productive ocean waters are green and turbid, teeming with ocean life. The major elements carbon (C), nitrogen (N), and phosphorus (P) are found in plant material in the molar ratios of 106 to 15 to 1 where they make up the principal biochemical plant components: sugars, fats, proteins, and nucleic acids. Living plants require these elements as dissolved inorganic nutrients for growth and reproduction. While C is plentiful (around 2 $mmol.kg^{-1}$ as noted above), N and P are in short supply in the ocean (rarely exceeding 45 and 3 $\mu mol.kg^{-1}$, respectively) and often become limiting for plant growth in stratified surface waters. In nearshore waters, however, excess nutrients can cause unchecked algal growth leading to eventual collapse of the populations and ensuing anoxia due to bacterial respiration, a condition known as eutrophication. Fertilizer in agricultural runoff, and human, domestic animal, and food industry waste are some of the major anthropogenic sources of nutrients to the aquatic environment. Habitat degradation with loss of aesthetic quality, massive fish kills, and harmful algal blooms are some of the consequence of anthropogenic eutrophication of coastal waters. Economic consequences can be severe. Sustained observations of these nutrient concentrations in nearshore waters provide forewarning of such events and assists in the identification and remediation of hotspots.

Inorganic dissolved phosphorus exists in solution in seawater for the most part as the bi-protonated phosphate ion ($H_2PO_4^-$) or its mono-protonated form (HPO_4^{2-}). These forms are collectively referred to as orthophosphate. Inorganic nitrogen on the other hand can exist in a number of chemical species ranging from the highly reduced ammonia/ammonium equilibrium couple (NH_3/NH_4^+) with a redox number of -III, through molecular nitrogen (N_2) with redox zero through several other forms including the laughing gas nitrous oxide (N_2O) redox +I, to the highly oxidized nitrite (redox +III) and nitrate (redox +V) ions. Of these, molecular nitrogen is the most abundant with around 400 $\mu mol.kg^{-1}$ in solution in surface waters. Molecular nitrogen is however largely unavailable to most taxonomic groups for biosynthesis due to the strength of the triple bond between atoms. This bond can only be attacked biologically by those organisms equipped with the nitrogenase enzyme, a few genera of bacteria and cyanobacteria. Thus, ammonium, nitrate and nitrate are the inorganic nitrogenous species most rapidly cycled in coastal marine waters. One more element, silicon (Si), although not a true metabolite, is traditionally classed among the essential nutrients due to its role in forming the skeletal material of diatoms as amorphous silica (SiO_2). In the ocean it is found mostly as dissolved silicic acid (H_4SiO_4) and is referred to in the literature as silicate. Iron, while limiting in some open ocean waters, is usually abundant in nearshore waters.

Traditional analytical procedures for characterization of these dissolved nutrients are performed through wet chemistry protocols during which reagents (added in excess) react to form colored products. The color intensity of the resulting solution is dependent on the analyte concentration which constitutes the reagent limiting to the reaction. The optical density or absorption (A) of the colored solution is measured by means of a spectrophotometer (an instrument equipped with a diffraction grating monochromator) or a simpler colorimeter (equipped with colored optical filters) against a set of standards of known concentration prepared in the laboratory including a zero standard known as the reagent blank. Laboratory protocols for wet chemistry analysis of reactive phosphorus and reactive silicate through the formation of colored molybdate complexes and of nitrite and nitrate through diazotization (with prior reduction of nitrate to nitrite) have changed little since their original descriptions. Strickland and Parsons (1972) provide authoritative accounts of the development and practice of these procedures. Limits of detection using the standard 10 cm cell are: 0.3 $\mu mol.l^{-1}$ for phosphate (reactive phosphorus), 0.1 $\mu mol.l^{-1}$

for reactive silicate, 0.01 μmol.l^{-1} for nitrite, and 0.05 μmol.l^{-1} for nitrite. Nitrate analysis through diazotization, as noted above, requires prior reduction to nitrite which is usually accomplished by passing the sample through a column packed with cadmium metal shavings coated with semi-colloidal copper. This column is extremely sensitive to oxidation and must be maintained submerged in ammonium chloride while not in use. Also repeated exposure to anoxic waters carrying sulfide ion will inactivate the column. The phenol/hypochlorite method described for ammonium analysis has been widely replaced with the more sensitive and less hazardous procedure of ortho-phthaldialdehyde OPA derivatization followed by fluorometric analysis (Holmes et al. 1999).

Many of these procedures have been automated and are now routinely performed in the laboratory using dedicated instruments known as flow-injection analyzers. These instruments propel aqueous samples and reagents with multichannel peristaltic pumps. Samples are automatically fed from carousel or linear array sample racks to injection valves where a small-volume sample plug is inserted into the carrier fluid flow. Method-specific capillary tubing manifolds allow injection of reagents to the carrier stream. Following transit through mixing coils, the resulting colored solutions are passed through flow-through colorimeters or fluorometers equipped with appropriate color filters. Analytical standards, either purchased or prepared in the laboratory at concentrations that bracket the expected sample concentrations, are analyzed concurrently for calibration. Preparation of standards and reagents in aqueous solution requires assurance of the absence of contamination with the target analyte in the dilution water or in the reagents. Prior to use, raw tap water is routinely filtered, distilled, and then passed through ion exchange columns to remove contaminants. Alternatively, for extremely exacting work where nutrient concentrations are very close to the method limit of detection, certified water may be purchased from chemical suppliers.

Autonomous Wet Chemistry Instruments

These technologies described above served as models for the current generation of autonomous in situ instruments. Several such chemical-laboratories-in-a-can are available on the market. Some manufacturers, responding to the need for conserving reagents in autonomous instruments, have opted to switch from sample injection as used in laboratory instruments to reagent injection since the instrument is after all immersed in the sample. Instruments incorporating microfluidic devices are now becoming available. In these miniaturized devices, fluid flow is along a network of microchannels etched directly on a chip rather than having capillary tubing transporting the sample. Instrument reliability is improved with the elimination of loose capillary tubing, end-fittings, and manifolds. Advantages of these low-volume devices are reduced reagent use and analysis time, high sample throughput, and the possibility for incorporating parallel analyses on multiple chips.

Spectroscopic Nutrient Measurements

An alternate avenue for nitrate analysis is afforded by the marked optical absorption peak of this ion in solution, an attractive analytical alternative since reagents need not be added. Commercial instruments based on this principle are available (Johnson and Coletti 2002). Interference of naturally occurring, optically active, dissolved inorganic bromide and bisulfide ions has been largely overcome by measurement of seawater absorption in 35 optical channels within the narrow wavelength band of 217–240 nm followed by multivariate linear regression. Optical opacity of glass at these wavelengths dictates the use of quartz optical windows. Further accuracy is achieved by accounting for instrument *dark intensity*; readings observed in the absence of irradiation (Martin et al. 2017). However, as noted below, absorption by colored dissolved organic matter (CDOM) is minimal in the red portion of the spectrum but increases exponential through the blue and UV causing substantial interference to UV nitrate determinations in waters with high CDOM content such as coastal and estuarine waters (Edwards et al. 2001; MacIntyre et al. 2009). Variability of the CDOM absorption spectrum as a result of both the molecular nature of the material and photodegradation suffered in transit has further hampered efforts to develop algorithms capable of discriminating the contribution of CDOM to the total absorption signature. Instrument accuracy is typically above 2 μM (Martin et al. 2017), making the method unusable for oligotrophic waters of low nutrient content.

The Alliance for Coastal Technologies recently (2017) announced the results of a Nutrient Sensor Challenge carried out under the auspices of a coalition of private, academic, and government organizations. Of 29 registered participants 5 were selected for the field phase of the challenge. It is encouraging to note that, while the winner was a conventional lab-in-a-can flow injection instrument, an honorable mention was extended to the manufacturers of a lab-on-a-chip microfluidic device. Through continued challenges, ACT seeks to promote the development of instruments providing precise and accurate phosphate and nitrate measurements and reliably surpassing a 3 month deployment threshold.

2.4 Electro-Optical Sensors for Measurement of Organic Matter in Seawater

2.4.1 *Colored Dissolved Organic Matter in Seawater*

Chemical Nature, Distribution, and Pertinence of CDOM to Coastal Ocean Observing

Dissolved organic matter (DOM) in natural waters may or may not interact with light in the UV/Vis spectrum. Analysis of DOM is largely performed in the laboratory using the amount of dissolved organic carbon (DOC) in seawater as an index of

DOM through high temperature combustion (e.g., Del Castillo et al. 1999). *Colored dissolved organic matter* (CDOM) is by definition the fraction of organic material dissolved in natural waters that absorb visible light; the remaining dissolved organic material is optically inactive and is mainly composed of small aliphatic molecules that absorb only in the ultraviolet if at all. Colored substances profoundly affect the penetration of light into natural waters bodies lending a tea-like tint to the water. *Black waters* containing high CDOM concentrations are common in river plumes and mangrove environments (Fig. 2.15).

Degradation of organic matter, particularly of terrestrial materials with high tannin contents, results in release to solution into coastal waters of globally significant quantities of these *chromophoric* (optically active) substances characterized by large molecular weight, polyphenolic content and variable chemical composition. In situ photochemical condensation reactions are recognized as a less intense oceanic source. These materials are commonly known as *humic* substances. Hansell and Carlson (2015) provide a detailed discussion of the composition, distribution, optical activity, and global biogeochemical significance of CDOM.

Terrestrial runoff, mainly from large rivers with high humic content, can modulate optical properties of near-surface waters over ocean areas extending thousands of kilometers from the river mouth. The *Coriolis parameter*, highest at high latitudes

Fig. 2.15 *Black waters* in the buoyant Orinoco River plume off the island of Grenada

and decreasing to zero at the equator, dictates differential lateral displacement of river plumes such that they disperse directly out into the ocean at very low latitudes but deviate strongly along the coast as latitude increases toward the right in the northern hemisphere and left in the southern. At higher latitudes, consequently, greater extensions of coastline come under the influence of river plumes. Terrestrial runoff in Arctic regions is now a significant source of CDOM to the ocean due to the anomalous melting of high peat-content permafrost.

CDOM is of practical concern since it limits the penetration of light into ocean waters affecting the availability of photosynthetically active radiation (400–700 nm). Thus, CDOM concentration limits the depth range of coastal vegetation such as seagrass and kelp beds and coral reefs. Conversely, CDOM may provide a UV screen to marine organisms residing close to the ocean surface reducing the harmful effects of UV light. Communications using optical means can be similarly hampered as will diver visibility. Since CDOM absorbs mostly in the blue to violet range where light penetration in seawater is greatest, the effect is particularly pertinent to these concerns.

Optical Characterization of CDOM

Absorption spectra of CDOM in the near-UV (about 320–400 nm) and visible (400–700 nm) bands conforms to hyperbolic parameterization with high absorption in the more energetic UV and violet-blue visible range and nonlinear decay toward the red extreme merging with water absorption peaks in the red band (Fig. 2.16).

Quantification of diagnostic absorption *peaks* (localized maxima) is negated due to their absence in the optical band. *Spectral slope*, the steepness of slope curvature of the parabolic CDOM absorption curve, increases through photo-oxidation and is

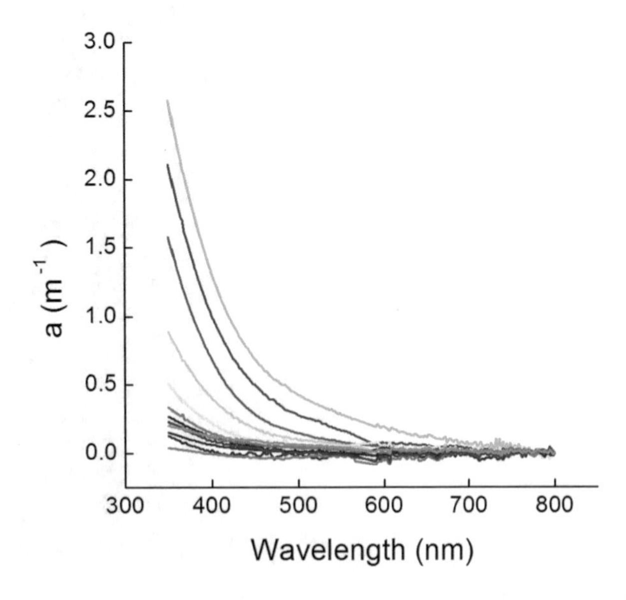

Fig. 2.16 Absorption spectra of surface waters in a gradient across the Orinoco River Plume in the Caribbean basin. CDOM absorption is greatest in the vicinity of the river delta around 10 °N and decreases along the gradient to 18 °N

used as a weathering parameter. Since the curves over a given spectral range adjust well to parabolic decay in a form akin to that of Eq. 2.1, logarithmic transformation yields the spectral slope parameter in units of $m^{-1}.nm^{-1}$. The contribution of CDOM to total DOC is variable depending on its chemical nature, the magnitude and periodicity of input to the coastal ocean and, due to its photolability, to the history of light exposure of the source material.

Spectrophotometric parameterization favors the lower UV bands where absorption is greatest and, despite more energetic solar irradiance, stratospheric ozone absorption significantly reduces radiant flux and thus the propensity for photodegradation. Parameters such as optical absorption at a specific wavelength or *spectral slope* of the absorption decay curve across a predetermined wavelength range serve for quantitative characterization. Absorbance at 300 nm of filtered samples (to remove particulate matter) is commonly used as an index of CDOM abundance (Del Castillo et al. 1999). However, for comparison to satellite ocean color instrument data, the 412 nm and 443 nm bands are also frequently used. Submersible multispectral (3–9 bands) and hyperspectral (400–750 nm) absorption and attenuation meters are commercially available, the latter taking readings in about 80 narrow spectral bands. However, these instruments are extremely sensitive to temperature, salinity, and fouling of the optical surface so their use is largely limited to research applications.

CDOM fluorescence offers a practical alternative for operational CDOM detection. High aromatic content of CDOM molecules imparts the capability for fluorescence under UV irradiance allowing a more robust approach to CDOM determination. Excitation/emission matrix fluorescence spectroscopy, whereby both excitation and emission (ex/em) are scanned across the UV-Vis spectrum using laboratory benchtop instruments, coupled to *parallel factor analyses*, a statistical technique for deconvolution of complex spectra, allows discrimination of discrete CDOM fluorophores. Simpler, single or multiple-band submersible ex/em instruments used for operational CDOM detection are widely available. Band choices, tuned to these fluorophores, range between 310–370 nm excitation and 450–470 nm emission. These instruments generally exhibit an orthogonal arrangement of the exciting radiation source and the emission sensing photodiode. Given the random molecular structure of humic substances, no single CDOM molecule exhibits the same fluorescence properties and no truly representative analytical standard exists. A common organic chemical calibration proxy is quinine sulfate with ex/em peaks at around 310 and 450 nm. Single design submersible fluorescence instruments now available may be configured for measurement of most of the fluorophores of interest to ocean observing including CDOM, here discussed, and petroleum hydrocarbons and chlorophyll, discussed below.

Remote satellite detection and quantification of CDOM in the Orinoco River Plume (Del Castillo et al. 1999) and the South Florida shelf (Müller-Karger et al. 2005) has been explored using empirical and semianalytical algorithms specific for CDOM retrieval. Instrument resolution currently limits operational applicability (Del Castillo 2005) to waters of very high CDOM content.

2.4.2 *Sensing and Tracking Petroleum Pollution in the Marine Environment*

Chemical Composition, Weathering, and Effects of Petroleum Pollution

Petroleum is a complex mixture of hydrocarbons that may be naturally released into the marine environment through sediment and coastal seeps. However, involuntary spills incurred during oil and gas exploration, and their transport and use, have vastly increased the volume of petroleum and its products entering marine water. Crude petroleum and its refined products are generally lighter than water, although a few heavier bitumen *nonfloating oils* may have a greater density than seawater (1.02–1.03 g.cm^{-3}) and will sink when released upon the ocean surface. Petroleum hydrocarbons are largely insoluble in water exhibiting bulk mass fractions ranging from well below a few micrograms per kg.

Most petroleum seeps and spills initially make their way to or remain on the surface creating slicks and subsequently undergo physical, chemical, and biological *weathering* changes that promote evaporation, dissolution, dispersion, entrainment of mineral particulates, and microbial degradation. Initial weathering losses are due to evaporation of the lighter fractions and moderate dissolution of the more soluble fractions. Extended weathering in moderate climatic conditions leads to the formation of tar balls which may eventually sink with the loss of lighter fractions and the added weight of biofouling organisms. High seas and winds on the other hand promote the formation of a persistent oil/water emulsion known as mousse (due to its similarity to the chocolate dessert).

Petroleum brings about decreased biological function, and death in the extreme, through various physical and chemical processes. Bird feathers and mammal pelts coated with oil lose their insulating properties leading in extreme to death from hypothermia. Aromatic hydrocarbons, the most soluble (or rather least insoluble) fraction of crude petroleum, and light derivatives such as gasoline are both toxic and mutagenic. Petroleum spills reaching the coastal zone lead to particularly adverse consequences to productive ecosystems such as marshes, mangroves, oyster beds, seagrass beds, and coral reefs. Persistence of petroleum that finds its way into these environments is shortened in environments subject to high wave energy (Evans et al. 2017) but can extend to decades in low energy environments such as mangrove ecosystems (Corredor et al. 1990). Further reference to *inputs, fate, and effects of oil in the sea* is available in two comprehensive *consensus study reports* of the US National Research Council (National Research Council 1985; Transportation Research Board and National Research Council 2003).

Oil spill response depends on timely identification for containment and treatment or, preferentially, removal. Since oil rapidly disperses in thin slicks upon the ocean surface, rapid confinement is of essence. For the reasons discussed above, however, regulatory agencies may prescribe dispersion of the product throughout the water column using chemically formulated *dispersant* to prevent spills from reaching the especially vulnerable coastal zone. This dispersion can affect planktonic organism due to toxicity of the aromatic fraction. Moreover, the dispersants currently available are themselves toxic, compounding the effects on planktonic communities.

Chemical and Optical Analysis of Petroleum in the Marine Environment

Definitive analysis of petroleum and its by-products in water, sediments, and organisms is today performed using coupled gas chromatography and mass spectrometry, complex techniques requiring compressed gases, high vacuum, and powerful magnetic fields. Chemical separation of most of the individual compounds is achieved by gas chromatography and their unequivocal identification and quantification by mass spectrometry. Extreme care is required in sample collection to avoid contamination and painstaking sample preparation is required prior to analysis. Nonetheless, the more toxic cyclic *aromatic* fractions of petroleum are optically active in the UV region and thus may be detected and quantified by optical means very similar to those described above for CDOM.

Optical Activity of Aromatic Compounds
Aromatic compounds derive their optical activity in the UV/Vis bands to delocalization of electrons into clouds of toroidal probability distribution to either side of the planar ring surface of a six member carbon ring, the so-called *delocalized pi* electron orbitals. Concatenation of adjacent rings further delocalizes the cloud shifting absorption to lower wavelengths encompassing the visible region.

Crude petroleum light absorption is greatest in the UV A, B, and C bands (200–400 nm) but strong also in the visible band and thus appears as brown or black to the eye (*black gold*). Aromatic hydrocarbons may contain different numbers of conjugated aromatic (benzene) rings such that optical absorption is spread out in the spectrum since each individual ring absorbs in concert with its neighboring electron clouds. The resulting UV/Vis absorption spectrum from about 300 nm conforms to a rather featureless exponential decay function as is the case for CDOM albeit incorporating small peaks attributed to optically active heavy metal chelates. In the absence of clear diagnostic absorption peaks, parameters such as specific absorption coefficients at key wavelengths and spectral slope within narrow bands (about 20 nm) provide objective numerical characterization.

Each component ring of a *polynuclear aromatic hydrocarbon* (PAH) constitutes a *fluorophore* with different excitation/emission properties responding to its immediate electron distribution cloud. Mathematical techniques are required to deconvolute the complex ex/em spectra resulting from the number of PAH fluorophores in a crude oil sample and their spectral overlap. Parallel factor analysis applied to data arising from instrumental excitation-emission matrix spectroscopy analysis, much as used for CDOM, enables discrimination and quantification of individual PAH components (Zhou et al. 2015). Less precise estimates of petroleum PAH dissolved and dispersed in seawater can be obtained using ex/em pairs tuned to abundant known aromatic compounds. Chrysene, for example, a fused four ring member PAH with ex/em maxima at 310 and 360, has long been used as a proxy standard (Atwood et al. 1987, 1987/1988). Methods available then required retrieval of water samples and extraction in hexane prior to fluorometric analysis using laboratory instruments. Today simple, submersible single optical band excitation/emission instruments equipped with LED photodiode emitters and borosilicate windows allow excitation down to about 350 nm in the near UV and wide emission bandpass filters (410–550 nm) irradiating photodiodes that allow monitoring for high concentrations of crude oils. As more capable UV LED emitters and UV transparent optical windows become available, more capable instruments should appear.

Remote Sensing of Petroleum Spills

Satellite spill detection is facilitated by the formation of buoyant ocean surface plumes whose dispersion may be imaged and tracked using optical instruments and synthetic aperture radar. Both depend on the dampening effect of oil on ocean waves in opposite fashion. Attenuated wave action reduces wave-induced refraction thus attenuating the *bottom-of-the-swimming-pool* effect perceived visually as webs of diminished or enhanced light. As a result, in areas where waves are dampened by oil slicks, images from visible light radiometers reflect a lower albedo due to absorption and scattering by seawater and its dissolved and suspended constituents. An important exception, and indeed a technique exploiting this same effect, is the exacerbation of *sunglint*, the mirror effect enhanced by the slick oil surface when the sun reflects directly onto the optical sensor. Radar microwaves on the other hand do not penetrate the ocean surface and are readily scattered by wave action returning a fraction of the radiation to the radar detector. Slick ocean surfaces reflect radiation more coherently away from the source unless the sensor is at direct nadir of the target slick.

The many variables described above forbid fully automated tracking of oil slicks by satellite remote sensing. Experienced human operators are still required to usefully interpret combined passive UV/Vis and active radar satellite imagery for spill detection and tracking.

2.5 Sensors for Biological Compounds and Processes: Chlorophyll, Accessory Pigments, and Photosynthetic Activity

Sustained instrumental observation of phytoplankton biomass, productivity, and community composition informing fisheries managers, health officials, and wildlife managers is now an integral component of coastal ocean observing systems. Phytoplankton, microscopic free-living photosynthetic cells and cell colonies suspended in the water column, constitute the bulk of the plant life in the ocean. A variety of mechanisms are used by phytoplankton to avoid sinking and thus remain in the upper photic layer of the ocean where they can actively photosynthesize. Phytoplankton constitutes the primary source of food for higher level marine organisms so phytoplankton biomass, community composition, and productivity rates constitute vital information for informed fishery management. Some phytoplankton which produce highly toxic metabolites cause the harmful algal bloom (HABs) known as red tides. These constitute a direct public health menace through respiration of aerosols and an indirect menace through consumption of contaminated shellfish. Shellfish fisheries are routinely closed under HAB threats. Animals may also be endangered. Recently for example die offs of sea lions along the west coast of the USA have been attributed to HAB poisoning. Public health officials consequently require continued up-to-date information on the presence and abundance of such organism in the plankton.

Since the principal photosynthetic pigment of plants, chlorophyll *a* (Chl *a*), interacts strongly with visible light, it can be measured in the field using optical techniques. Given this property and despite the variable content of Chl *a* relative to other more suitable chemical proxies such as carbon, its concentration in the water column is often used as a chemical proxy for estimating plant biomass expressed as Chl *a* mass per unit volume ($\mu g.l^{-1}$ or its equivalent $mg.m^{-3}$). In addition to Chl *a*, commonly the principal component, the plant photosynthetic apparatus comprises numerous accessory pigments including various other chlorophylls (*b, c*, c1, c2, c3) plus xanthophylls and carotenes. These compounds are sparingly soluble in water if at all and in nature are found packaged in hydrophobic environments within the chloroplast or cell structure. Photosynthetic bacterioplankton (known colloquially as *blue-green algae*) additionally contain the water-soluble photosynthetic protein pigments phycocyanin and phycoerythrin containing phycobillin chromophores.

2.5.1 *In Vitro/In Vivo* Chlorophyll Fluorometry

Chlorophyll *a* concentration has been traditionally determined spectroscopically following seawater filtration, sample grinding in acetone or other organic solvents and extract clarification through centrifugation or filtration. During this procedure, pigments are removed from their cell/chloroplast packaging and dissolved in an organic matrix. Chlorophylls *a, b*, and *c*, having similar chemical structure, have similar but clearly discernible absorption spectra with peaks in the blue and red spectral bands and a prominent valley in the green, hence the green color of most plants. Thus, judiciously chosen diagnostic measurements at the absorption peaks of each of these pigments allowed the development of *trichromatic equations* that served to solve for the three chlorophylls (and thus Chl *a*) from spectrophotometric measurements of bulk pigment extracts (Strickland and Parsons 1972). Conversely to the chlorophylls, xanthophylls and carotenes, accessory pigments which contribute to the bulk light absorption of the clarified extract, absorb in the green and blue spectral bands and appear yellow, orange, or red to the eye. Pigment extracts can further be separated individually in the laboratory by high performance liquid chromatography, optically scanned and individually identified from their unique optical spectra and chromatographic retention properties. However, with over 80 currently recognized chemically distinct photosynthetic pigments, high operational costs, and low sample throughput, HPLC remains a research tool. Such laborious extractive techniques are in general unsuitable for operational ocean observing but are required for calibration of the bio-optical instruments described below.

Chlorophylls exhibit fluorescence in the visible band: when Chl *a* extract is irradiated with blue light (460–470 nm), blood red fluorescence (685–695 nm) is visually observed. Fluorescence analysis can be over two orders of magnitude more sensitive than spectroscopic analysis and is much more robust to confounding suspended material in the sample. Given its greater versatility, fluorescent analysis is favored over the spectrophotometric approach and has been exploited for Chl *a* estimation in oceanography since suitable instruments were developed in the 1950s. Fluorescence analysis is similar to the photometric procedure; a clarified acetone extract is poured in an appropriate cuvette and placed in the instrument. In bench top instruments, a light source is conditioned with a blue narrow band (colored) filter and aimed to excite the sample. A photomultiplier, usually mounted at right angles to the light source and shielded by a narrow-band red filter receives the emitted light and rejects residual excitation light. Both Chl *b* and Chl *c* also fluoresce but appropriate choice of the narrow band filter removes this interference. Carotenoids do not fluoresce. Phycoerythrin fluoresces yellow (575 nm) under green excitation (525 nm) and is widely used to discriminate between the eukaryotic phytoplankton and the blue-green cyanobacteria. However, phycoerythrin is a water-soluble pigment and must be subject to aqueous extraction.

Extractive analysis as described above is not easily adapted to autonomous measurement given the necessity for sample disruption, organic extraction, and clarification of the extract. In vivo fluorescence analysis is on the other hand feasible and widely used despite its many shortcomings. Photosynthesis transforms light energy into chemical bonds. Thus, any light absorbed but then reemitted by a live cell under illumination is light not absorbed in the photosynthetic process. In nutrient-rich near-surface environments under high irradiance, highly productive cells in full growth fluoresce less. On the contrary, the photosynthetic process of nutrient starved cells is inefficient and, in these cells, light not used is fluoresced. In stratified waters, cells in the deep chlorophyll maximum at depths between 30 and 150 m are strategically situated close to the nutracline ensuring an ample nutrient source but are light limited so they tend to have a high Chl *a*:C ratio. These cells fluoresce proportionally less when excited than near-surface cells do. Moreover, while extractive photometric analysis deals with a homogeneously dissolved pigment in an optically clear solution, photosynthetic pigments in vivo are sequestered in individual cells, some in filamentous colonies, and, in eukaryotes at least, are further packaged within cell chloroplasts. The medium is optically turbid and *self-shading* is substantial, both within and between cells. Chloroplast structure moreover changes throughout the day, expanding and contracting with the irradiance cycle. These variables further contribute to the uncertainty of in vivo Chl *a* measurements.

Despite the shortcomings described, a large number of submersible *chlorophyll fluorometers* are in use today and a wide selection of instruments is available.

Such instruments must be calibrated in the field in order to reflect the particular environmental conditions of the phytoplankton community and must be recalibrated as conditions change. These instruments, described below, are a subset of the now common submersible fluorometer family of instruments used for the various other analyses described above. Many designs feature a common platform where the light source and filters are substituted to specifications as the application may require.

Early submersible fluorometers adaptable to modern CTD/carrousel arrays incorporated traditional cast iron optical benches with various lenses, xenon lamps, and photomultiplier detectors housed in a cylindrical pressure hull. Power requirements were substantial, severely limiting autonomous operation. Instruments in use today for the most part use LED sources and photodiode detectors, vastly reducing bulk, weight, and power consumption. Single optical excitation and emission bands are selected using optical filters. Various instrument architectures are available for pumped or unpumped operation. Most common is an external cell array with orthogonal light source and detector aimed at a small volume of the seawater medium. Some instruments mount two opposed light sources. External optical arrays may allow installation of optical head housings with inlet and outlet suitable for pumped operation. A more innovative solution is an axial flow-through optical pipe essentially transporting the optical head into the instrument housing. These instruments operate at low power (7–20 V DC) drawing less than 100 mA (Fig. 2.17).

For shipboard or mooring applications, data generated by fluorometers and other instruments are usually coupled to the primary onboard environmental data systems; CTDs or thermosalinographs, together with the now common oxygen, pH, and pCO_2 sensor data. Ancillary data from these instruments is appended to the CTD and GPS data streams for telemetry.

2.5.2 *Automated Sell Sorting, Counting, and Bio-optical Characterization*

A number of phytoplankton species including some diatoms and, most notably, dinoflagellates are capable of producing highly toxic water-soluble compounds that affect marine as well as terrestrial organisms. Human consumption of contaminated shellfish causes poisoning leading to severe gastric and neurological distress and resulting, in extreme, in paralysis and eventual death. Even inhalation of aerosols of ocean foam containing the organisms or their detritus can result in illness. Their proliferation in coastal environments has been attributed to anthropogenic eutrophication that, under appropriate environmental circumstances, promotes their unchecked growth resulting as *harmful algal blooms* (HABs). Chlorophyll biomass determination provides only an indication of algal biomass. Fortunately, automated means of determining the composition of phytoplankton assemblages and a mature understanding of the role of turbulence and nutrient supplies on their temporal evolution are now beginning to allow an operational approach to forecasting HABs.

Fig. 2.17 Single ex/em band axial flow submersible fluorometer

Automated *flow cytometers* or cell *sorters* were initially developed for clinical applications, namely sorting and counting blood cells. Pioneering instruments quantified individual cells by monitoring electrical resistance across a small orifice through which the liquid stream containing the cells was passed. Since each cell would obstruct the orifice, electrical conductivity depended on the rate of flow and the size of the individual cell. These capabilities were rapidly adopted for oceanographic use and a number of automated cell sorting instrumentation based on optical detection are now available.

At the heart of modern environmental flow cytometry is the ability to funnel sample cells in the seawater medium sheathed in a fluid *carrier* stream to achieve single cell passage through the sensing volume. Multiple wavelength laser probes allow both fluorescence and scattering analysis of individual cells (Dubelaar et al. 1999; Dubelaar and Jonker 2000). A compact system incorporating two excitation lasers, multiple fluorescence detectors, and multiple scatter sensors is commercially available in configurations for either laboratory operation or for field deployment mounted as a submersible profiling unit or in a buoy package. Dugenne et al. (2015) have demonstrated the feasibility of detecting and following the evolution of a HAB using a buoy-mounted instrument but operational long-term deployment of such automated flow cytometers is still hampered by the complexity of the instrument and its relatively short autonomy.

The benchtop instrument originally described by Sieracki et al. (1998) provides an alternative to laser flow cytometry. A seawater stream is passed through a flat optical chamber where digital images and wide-band fluorescence signals (435–470 nm excitation; 520–700 nm emission) are recorded. Simultaneous imaging and fluorescence peak identification allows selecting particles with similar size, aspect ratio, and fluorescence properties. Digital imaging allows automated shape recognition so the associated software systems may be *trained* to recognize and quantify individual species. These instruments are suitable for laboratory or shipboard work and have successfully monitored HABs bloom evolution (Buskey and Hyatt 2006) (Fig. 2.18).

2.5.3 Remote Sensing of Photosynthetic Pigments

Operational coastal observing applications of satellite-derived ocean color data focus on the occurrence, distribution, and fate of algal blooms which may be health hazards or may constitute navigation and aesthetic nuisances. Blondeau-Patissierre et al. (2014) provide a detailed and up-to-date review of the use of satellite ocean color data for detection mapping and analysis of phytoplankton blooms. They note the severe problems yet faced for accurate retrievals in the coastal zone and highlight the value of coupling ocean color data to additional environmental data sets, statistical methods, and numerical models.

Several substances either dissolved or dispersed in the water column interact with visible light and the adjacent IR and UV bands allowing their detection in near-surface waters using satellite-borne radiometers. Phytoplankton photosynthetic pigments, dissolved organic matter, mineral phytoplankton inclusions, and suspended nonphotosynthetic particulate matter are among the members of this photoactive class of substances. Detection of photosynthetic pigments and dissolved organic matter is discussed below. Biologically generated calcium carbonate inclusions common to the *coccolithophores* (planktonic single celled eukaryotic algae) are readily apparent in images of massive blooms occurring in surface waters at temperate and polar latitudes. Suspended nonphotosynthetic particulate matter is of limited

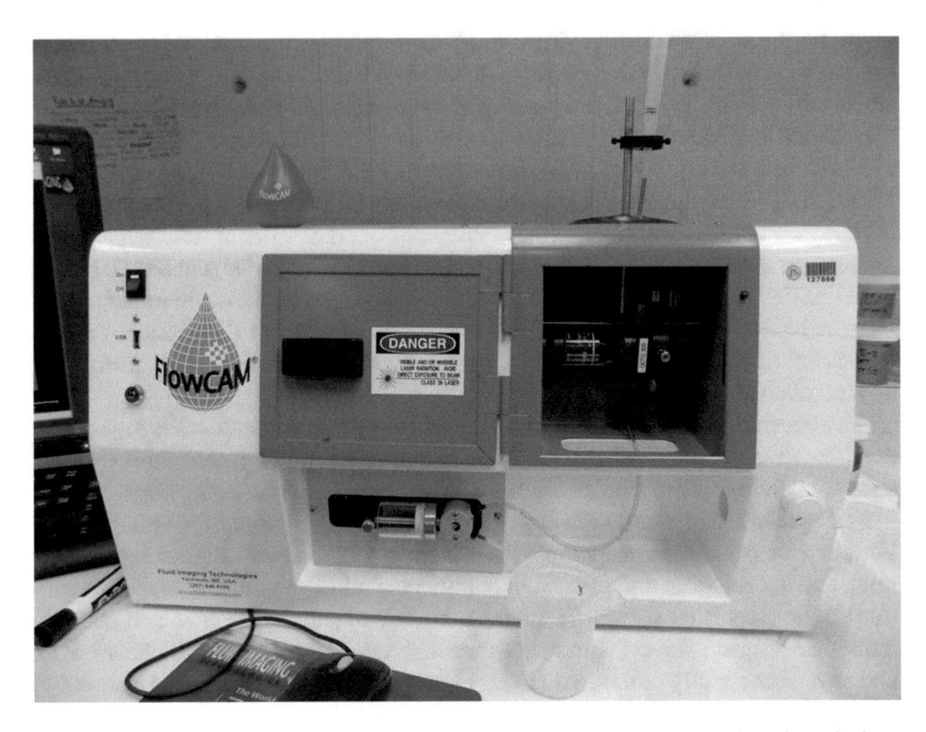

Fig. 2.18 Benchtop flow-through imaging and fluorescence instrument for phytoplankton characterization

concern since it is, for the most part, limited to the mouths of turbid rivers and estuaries. Since calibration of ocean color instruments in space can be performed directly by observation of the lunar disk, allowing evaluation of instrument degradation. However, specific calibration for complex multiple and variable photophores such as phytoplankton and CDOM cannot be performed directly. *Vicarious calibration* is performed indirectly through observation at sea of the materials interacting with light, concurrent with satellite overpasses (Clarke and Mobey 2002; Brown et al. 2007).

Space-based synoptic observations of near-surface ocean color began with the launching of the Coastal Zone Color Scanner (CZCS) aboard the polar orbiting Nimbus 7 weather satellite in 1978. The instrument was equipped with a rotating plane mirror covering a swath of 1556 km. Light was fed through a reflector telescope and a beam splitter to a polychromator which in turn irradiated five silicon photodiodes. For surface Chl *a* estimation, algorithms based on *band ratios* were developed to calculate the ratio of *water leaving radiance* (L_W) at 443 and 670 nm (where chlorophyll absorbs strongly) to the so-called *chlorophyll hinge point* at 550 nm where absorption is minimal (and hence the *greenness* of chlorophyll). Pixel resolution was 0.825 km centered at nadir. CZCS operated for eight consecutive years and proved the conceptual feasibility of satellite ocean color observing.

Subsequent to the CZCS a number of ocean color sensors have been placed in orbit with increasingly sophisticated capacity through the addition of more IR, visible, and UV sensor bands, improved calibration, and refined algorithms. To date, the following instruments have returned imagery: Ocean Color and Temperatures Sensor (OCTS 1996–1997), Sea Viewing Wide Field-of-View Sensor (SeaWiFS-1997–2010), OCM-1, Moderate-Resolution Imaging Spectroradiometer (MODIS – Terra 2000–present), MEdium Resolution Imaging Spectrometer (MERIS 2002–2012), MODIS-Aqua (2004–present), Visible Infrared Imaging Radiometer Suite (VIIRS 2011–present). A number of other US, Japan, EU, and Indian missions have recently commenced or are in cue for launch. Optical signatures of mesoscale structures of about 200 km such as river plumes and ocean eddies can be effectively characterized and tracked at a regional scale (Müller-Karger et al. 1989; Hu et al. 2004; Corredor et al. 2004).

Satellites bearing ocean color instruments in a sun-synchronous near-polar orbit at about 700–800 km altitude allow global coverage at a frequency of 1–2 days. Ocean color imagery products with resolutions to 1 km rendered as false-color Chl *a* are widely available through various government, academic, nonprofit organizations, and private enterprise. MODIS ocean color data is available directly through the appropriate NASA portal in many data levels, from L1, the raw band data to L3, where data is georeferenced, corrected for atmospheric effects, and rendered through appropriate algorithms to represent surface distribution of various parameters such as Chl *a* and CDOM. In the USA, most RCOOSs serve regional ocean color data on their web sites.

While the quasi-synoptic nature of the satellite images is invaluable, several drawbacks to operational remote sensing of phytoplankton must be considered. Atmospheric effects are cardinal since less than 20% of radiance reaching the satellite radiometer is water leaving radiance (L_W). In the case of deep blue offshore waters with sparse phytoplankton communities, L_W can be less than 5% (Müller-Karger et al. 2005). Clouds are opaque to visible and near-infrared radiation (Vis/IR) and are commonly masked out in imagery products. Nevertheless, cloud edges, extremely sparse clouds, aerosols, and dust can elude masking algorithms causing unacceptable interference. Atmospheric corrections including cloud and aerosol masks on the scale of hundreds of kilometers for regional images of near-surface optically active substances such as phytoplankton can reject data to the point of resulting in no useful image at all for a given day pass. Time averaged images of multiple satellite passes over days to weeks provide increasingly coherent but increasingly blurred pictures of optical sea surface properties as currents, eddies, and filaments transport and disperse surface waters.

NASA empirical Chl *a* retrieval algorithms use band ratios of peaks and valleys in the remote sensing reflectance (R_{rs}) spectrum. MODIS AQUA–derived Chl *a* products are currently computed using the (MODIS)/Aqua Ocean Chlorophyll three-band (OC3) or four-band (OC4) algorithms involving exponential polynomial expressions for the ratio of R_{rs} at 443 or 488 nm as the diagnostic peaks to R_{rs} at

551 nm as the ratio denominator (OC3), or the ratio of R_{rs} at either 443, 490, or 510 to R_{rs} at 555 nm (OC4) (O'Reilly et al. 1998; Werdell et al. 2009). In both cases, the ratio denominator is known as a *hinge point* since it is located at a wavelength in the spectrum where chlorophyll absorbs little and hence reflects green light. Various weighting mechanisms are used to select the ratio used such as choosing the highest for any pixel, or setting cutoffs to avoid natural spectral reflectance red-shifts at higher Chl *a*.

These algorithms are successfully applicable to waters where absorption or scattering by materials different from phytoplankton is minimal, but such conditions are not always met in coastal waters where suspended material and colored dissolved organic matter (CDOM) can play a large part in determining the effective R_{rs} spectrum. The exponential decay spectrum of CDOM, absorbing strongly in the blue but minimally in the red, can lead these simple algorithms astray in areas of high CDOM content. Since CDOM absorbs most strongly in the blue, large discrepancies between satellite-retrieved Chl *a* values and in situ measurements result when estimating Chl *a* using algorithms dependent on blue to green band ratios (Del Castillo 2005; Müller-Karger et al. 2005).

Spectral band-difference algorithms can allow accurate visualization of dense phytoplankton blooms. These differ from the band-ratio family in that baselines extending from the low reflectance PAR region across the high reflectance near infrared (NIR) into the short wave infrared (SWIR) allow direct estimation of the NIR peak of interest. Such approaches have been exploited successfully for detection of various algal blooms of practical concern. The fluorescence line height (FLH) algorithm (667, 678, and 746 nm) exploits in vivo fluorescence at 678 nm while the floating algae index (FAI), operating at 667, 859, and 1240 nm, and its modifications exploit the so-called red-edge across the red-NIR spectrum (Blondeau-Pattissier et al. 2014).

Extensive mats of floating *Sargassum sp*. appearing of late in the Caribbean have become a nuisance to nearshore marine operations as these rafts become entrained along beaches and harbors. Mats extending out tens of meters from the coast along windward shorelines particularly hamper small boat operations impacting tourism and fishing. *Sargassum* blooms have been successfully tracked and operational products are available at scales relevant to stakeholders: (http://www.caricoos.org/oceans/observation/modis_aqua/ECARIBE/afai Accessed 8/25/2017). Detection depends on a *floating algae index* across the *vegetation red edge* using the baseline between 667 and 1240 nm (or 1640) and the diagnostic peak at 859 nm where floating algal biomass is highly reflective (https://eos.org/features/sargassum-watch-warns-of-incoming-seaweed).

Specific products for mapping harmful algal blooms via satellite imagery use the MODIS and VIIRS chlorophyll data products to inform statistical models augmented by periodic sample collection as discussed below in Chap. 6.

References

Atwood DK, Kinard WF, Barcelona MJ, Johnson CE. Comparison of Polarographic electrode and Winkler titration determinations of dissolved oxygen in oceanographic samples. Deep-Sea Res. 1977;24(3):311–3.

Atwood DK, Burton FJ, Corredor JE, Harvey G, Mata-Jimenez A, Vasquez-Botello A, Wade B. Results of the CARIPOL petroleum pollution monitoring project in the Wider Caribbean. Mar Poll Bull. 1987;18:540–8.

Atwood DK, Burton FJ, Corredor JE, Harvey G, Mata-Jimenez A, Vasquez-Botello A, Wade B. Petroleum pollution in the Caribbean. Oceanus. 1987/1988;30:25–32.

Baringer MO, Larsen JC. Sixteen years of Florida current transport at 27°N. Geophys Res Lett. 2001;28(16):3179–82.

Barrick DE, Evans MW, Weber BL. Ocean surface currents mapped by radar. Science. 1977;198:138–44.

Barton A, Hales B, Waldbusser GG, Langdon C, Feely RA. The Pacific oyster, *Crassostrea gigas*, shows negative correlation to naturally elevated carbon dioxide levels: implications for near-term ocean acidification effects. Limnol Oceanogr. 2012;57(3):698–710.

Blondeau-Pattissier D, et al. A review of ocean color remote sensing methods and statistical techniques for the detection, mapping and analysis of phytoplankton blooms in coastal and open oceans. Prog Oceanogr. 2014;123:123–44.

Branham CW, Murphy DJ, Walsh ID. Reliably measuring pH in the ocean. Int Ocean Systems 2017; 21(5) http://www.intoceansys.co.uk/articles-detail.php?iss=0000000062&acl=0000000569. Accessed 4/28/2018.

Brown SW, Flora SJ, Feinholz ME, Yarbrough MA, Houlihan T, Peters D, Kim YS, Mueller J, Johnson BC, Clark DK. The Marine Optical BuoY (MOBY) radiometric calibration and uncertainty budget for ocean color satellite sensor vicarious calibration. Proc SPIE Optics Photonics Sensors Syst Next Generation Satellites XI. 2007;6744:67441M.

Bureau International des Poids et Mesures. The international system of units (SI). 8th ed. Organisation intergouvernementale de la convention du Mètre Paris: 2006. ISBN 92-822-2213-6.

Buskey EJ, Hyatt CJ. Use of the FlowCAM for semi-automated recognition and enumeration of red tide cells (Karenia brevis) in natural plankton samples. Harmful Algae. 2006;5(6):685–92.

Caldeira K. What corals are dying to tell us about CO_2 and ocean acidification. Oceanography. 2007;20:188–95.

Caldeira K, Wickett ME. Anthropogenic carbon and ocean pH. Nature. 2003;425:365.

Chatterjee A, Gierach MM, Sutton AJ, Feely RA, Crisp D, Eldering A, Gunson MR, O'Dell CW, Stephens BB, Schimel DS. Influence of El Niño on CO_2 over the tropical Pacific Ocean: findings from NASA's OCO-2 mission. Science. 2017;358(6360):190.

Clark DK, et al. MOBY, a radiometric buoy for performance monitoring and vicarious calibration of Satellite Ocean color sensors: measurement and data analysis protocols. In: Muller JL, Fargion GS, editors. Ocean optics protocols for satellite ocean color sensor validation, revision 3. Greenbelt: National Aeronautics and Space Administration, Goddard Space Flight Center; 2002. Volume 2 NASA/TMm2002–210004/Rev3-Vol2.

Clayton TD, Byrne RH. Spectrophotometric seawater pH measurements: Total hydrogen ion concentration scale calibration of m-cresol purple and at-sea results. Deep-Sea Res I. 1993;40:2115–29.

Coppola L, Salvetat F, Delauney L, Machoczek D, Karstensen J, Sparnocchia S, Thierry V, Hydes D, Haller M, Nair R, Lefevre D. White paper on dissolved oxygen measurements: scientific needs and sensors accuracy. JERICO report (EU FP7 project, grant agreement no: 262584). 2013.; http://www.jerico-ri.eu/download/filebase/White%20paper%20DO_final%20-copyright.pdf. Accessed 4/28/2018.

Corredor JE. Development and propagation of internal waves in the Mona Passage. Sea Tech. 2008;49(10):48–50.

Corredor JE, Morell JM, Del Castillo C. Persistence of spilled oil in a tropical intertidal environment. Mar Poll Bull. 1990;21:385–8.

Corredor JE, Morell JM, López JM, Capella JE, Armstrong RA. Cyclonic eddy entrains Orinoco River plume. EOS Trans Am Geophys Union. 2004;85(20):197, 201–2.
Corredor JE, Amador A, Canals M, Rivera S, Capella JE, Morell JM, Glenn S, Handel E, Rivera E, Roarty H. Optimizing and validating high frequency radar surface current measurements in the mona passage. Mar Technol Soc J. 2011;45(3):49–58.
Cyronak, Tyler, Karl Schulz and Paul Jokiel. The Omega myth: what really drives lower calcification rates in an acidifying ocean ICES J Mar Sci 2015 DOI: https://doi.org/10.1093/icesjms/fsv091. 5 p.
DeGrandpre MD, Hammar TR, Smith SP, Sayles FL. In situ measurements of seawater pCO2. Limnol Oceanogr. 1995;40:969–75. https://doi.org/10.4319/lo.1995.40.5.0969.
Del Castillo CE. Remote sensing of organic matter in coastal waters. In: Miller RL, Del Castillo CE, McKee BA, editors. Remote sensing of coastal aquatic environments. Dordrecht: Springer; 2005. p. 157–79.
Del Castillo CE, Coble PG, Morell JM, López JM, Corredor JE. Analysis of optical properties of the Orinoco River by absorption and fluorescence spectroscopy: changes in optical properties of the dissolved organic matter. Mar Chem. 1999;66:35–51.
Dickson AG. The carbon dioxide system in seawater: equilibrium chemistry and measurements. In: Riebesell U, Fabry VJ, Hansson L, Gattuso J-P, editors. Guide to best practices for ocean acidification research and data reporting, vol. 260. Luxembourg: Publications Office of the European Union; 2010.
Dore JE, Lukas R, Sadler DW, Church MJ, Karl DM. Physical and biogeochemical modulation of ocean acidification in the central North Pacific. Proc Natl Acad Sci. 2009;106(30):12235–40.
Dubelaar GBJ, Jonker RR. Flow cytometry as a tool for the study of phytoplankton. Sci Mar. 2000;64(2):135–56.
Dubelaar GBJ, Gerritzen PL, Beeker AER, Jonker RR, Tangen K. Design and first results of CytoBuoy: a wireless flow cytometer for in situ analysis of marine and fresh waters. Cytometry. 1999;37:247–54.
Dugene M, Thyssen M, Garcia N, Mayot N, Bernard G, Grégori G. Monitoring of a potential harmful algal species in the berre lagoon by ated in situ flow cytometry. In: Ceccaldi HJ, Hénocque Y, Koike Y, Komatsu T, Stora G, Tusseau-Vuillemin MH, editors. Marine productivity: ations and resilience of socio-ecosystems. Cham: Springer International Publishing; 2015. p. 117–27.
Edwards AC, Hooda PS, Cook Y. Determination of nitrate in water containing dissolved organic carbon by ultraviolet spectroscopy. Int J Environ Anal Chem. 2001;80(1):49.
Eldering A, Chris WO, apos; Dell, Wennberg PO, Crisp D, Michael R, Gunson CV, Avis C, Braverman A, Castano R, Chang A, Chapsky L, Cheng C, Connor B, Dang L, Doran G, Fisher B, Frankenberg C, Fu D, Granat R, Hobbs J, Richard A, Lee M, Mandrake L, McDuffie J, Charles E, Miller VM, Natraj V, Denis O, apos, Brien GB, Osterman FO, Vivienne H, Payne HR, Pollock IP, Coleen M, Roehl RR, Schwandner F, Smyth M, Tang V, Taylor TE, Cathy TO, Wunch D, Yoshimizu J. The Orbiting Carbon Observatory- 2: first 18Â months of science data products. Atmos Meas Tech. 2017;10(2):549–63.
Emery WJ, Wick GA, Schleussel P. Chapter 10. Skin and bulk sea surface temperatures: satellite measurements and corrections. In: Ikeda M, Dobson FW, editors. Oceanographic applications of remote sensing. Boca Raton: CRC Press; 1995. p. 145–65.
Evans M, Liu J, Bacosa H, Rosenheim BE, Liua Z. Petroleum hydrocarbon persistence following the deepwater horizon oil spill as a function of shoreline energy. Mar Pollut Bull. 2017;115:47–56.
Feely RA, Sabine CL, Takahashi T, Wanninkhof R. Uptake and storage of carbon dioxide in the ocean: the global CO2 survey: reference materials for oceanic CO_2 measurements. Oceanography. 2001;14(4):18–32.
Feely RA, Sabine CL, Hernandez-Ayon JM, Ianson D, Hales B. Evidence for upwelling of corrosive "Acidified" water onto the continental shelf. Science. 2008;320(5882):1490–2.
Fielding S. The biological validation of ADCP acoustic backscatter through direct comparison with net samples and model predictions based on acoustic-scattering models. ICES J Mar Sci. 2004;61(2):184–200.
Grey SEC, DeGrandpre MD, Langdon C, Corredor JE. Short-term and seasonal pH, pCO_2 and saturation state variability in a coral-reef ecosystem. Global Biogeochem Cycles. 2012;26:GB3012. https://doi.org/10.1029/2011GB004114.

Gurgel KW, Antonischki G, Essen HH, Schlick T. Wellen Radar (WERA): a new ground wave HF radar for remote sensing. Coast Eng. 1999;37:219–34.

Hansell DA, Carlson CA, editors. Biogeochemistry of marine dissolved organic matter. 2nd ed. London: Academic; 2015. 693 pp.

Holliday NP, Yelland MJ, Pascal R, Swail VR, Taylor PK, Griffiths CR, Kent E. Were extreme waves in the Rockall trough the largest ever recorded? Geophys Res Lett. 2006;33(5):L05613.

Holmes RM, et al. A simple and precise method for measuring ammonium in marine and freshwater ecosystems. Can J Fish Aquat Sci. 1999;56:1801–8.

Hu C, Montgomery ET, Schmitt RW, Müller-Karger FE. The dispersal of the Amazon and Orinoco River water in the tropical Atlantic and Caribbean Sea: observation from space and S-PALACE floats. Deep-Sea Res II Top Stud Oceanogr. 2004;51:1151–71. https://doi.org/10.1016/j.dsr2.2004.04.001.

Johnson KS, Coletti LJ. In situ ultraviolet spectrophotometry for high resolution and long-term monitoring of nitrate, bromide and bisulfide in the ocean. Deep Sea Res I. 2002;49:1291–305.

Johnson KS, Jannasch HW, Coletti LJ, Elrod VA, Martz TR, Takeshita Y, Carlson RJ, Connery JG. Deep-Sea DuraFET: a pressure tolerant pH sensor designed for global sensor networks. Anal Chem. 2016;88(6):3249–56. https://doi.org/10.1021/acs.analchem.5b04653.

Kirk JTO. Light and photosynthesis in aquatic ecosystems. London/New York: Cambridge University Press; 1994.

Lagerloef G. Satellite mission Monitors Ocean surface salinity. EOS Trans Am Geophys Union. 2012;93(2519):233–40.

Langdon C, Takahashi T, Sweeney C, Chipman D, Goddard J, Marubini F, Aceves H, Barnett H, Atkinson MJ. Effect of calcium carbonate saturation state on the calcification rate of an experimental coral reef. Global Biogeochem Cycles. 2000;14:639–54.

Lewis EL. The practical salinity scale 1978 and the international equation of state of seawater 1980. IEEE J Ocean Eng. 1980;OE-5(1):3–8.

Liu Y, et al. Development of Chinese carbon dioxide satellite (TanSat). Vienna: EGU General Assembly; 2013. p. 157–89.

MacIntyre G, Plache B, Lewis MR, Andrea J, Feener S, McLean SD, Johnson KS, Coletti LJ, Jannasch HW. ISUS/SUNA nitrate measurements in networked ocean observing systems. http://www.stccmop.org/files/ISUS-SUNA-Nitrate-Measurment-White-Paper.pdf. Accessed 9/11/2017; 2009.

Marsh HW. Underwater sound and instrumentation. In: Myers JJ, Holm CH, McAllister RF, editors. Handbook of ocean and underwater engineering. Copyright by North American Rockwell Corporation. New York: McGraw Hill; 1969. pp. 3–3 to 3–20.

Martin KI, Walsh ID, Branham CW. Measuring nitrate in Puget sound using optical sensors. Mar Tech Oct. 2017;58:10–113.

Martz TR, Carr JJ, French CR, DeGrandpre MD. A submersible autonomous sensor for spectrophotometric pH measurements of natural waters. Anal Chem. 2003;75(8):1844–50. https://doi.org/10.1021/ac0205681.

Martz TR, Connery JG, Johnson KS. Testing the Honeywell Durafet® for seawater pH applications. Limnol Oceanogr Methods. 2010;8:172–84.

McDougall TJ, Jackett DR, Millero FJ, Pawlowicz R, Barker PM. A global algorithm for estimating absolute salinity. Ocean Sci. 2012;8:1123–34. https://doi.org/10.5194/os-8-1123-2012. www.ocean-sci.net/8/1123/2012/.

Millero FJ. Chemical oceanography. 4th ed. Boca Raton: CRC Press Taylor & Francis Group; 2013. 571 p. ISBN 9788-1-4665-1249-8.

Millero FJ, Pierrot D, Lea K, Wanninkhof R, Feely R, Sabine CL, Key RM, Takahashi T. Dissociation constants for carbonic acid determined from field measurements. Deep Sea Res I. 2002;49:1705–23.

Mills A. Optical oxygen sensors utilizing the luminescence of platinum metals complexes. Platinum Metals Rev. 1997;41(3):115–27.

Müller-Karger FE, McClain CR, Fisher TR, Esaias WE, Varela R. Pigment distribution in the Caribbean Sea: observations from space. Prog Oceanogr. 1989;23:23–64.

Müller-Karger FE, Hu C, Andréfouët S, Varela R, Thunell R. The color of the coastal ocean and applications in the solution of research and management problems. In: Miller RL, et al., editors. Remote sensing of coastal aquatic environments. Dordrecht: Springer; 2005. p. 102–27.

National Research Council. Oil in the sea: inputs, fates, and effects. Washington, DC: The National Academies Press; 1985. https://doi.org/10.17226/314.

NOAA. Edwing R, Next generation water level measurement system NGWLMS site design, preparation, and installation manual. Rockville; 1991. pp. 213.

O'Reilly JE, Maritorena S, Mitchell BG, Siegel DA, Carder KL, Garver SA, Kahru M, McClain C. Ocean color algorithms for SeaWiFS. J Geophys Res. 1998;103:24,937–53. https://doi.org/10.1029/98JC02160.

Orr JC, Epitalon JM, Gattuso JP. Comparison of ten packages that compute ocean carbonate chemistry. Biogeosciences. 2015;12:1483–510. https://doi.org/10.5194/bg-12-1483-2015. www.biogeosciences.net/12/1483/2015/.

Paduan JD, Washburn L. High-frequency radar observations of ocean surface currents. Annu Rev Mar Sci. 2013;5:115–36.

Park, J, Heitsenrether R, Sweet WV. Water level and wave height estimates at NOAA . Tide stations from acoustic and microwave sensors. Silver Spring: NOAA technical report NOS CO-OPS 075. p. 41, 2014.

Peng G, Garra Z, Halliwell GR, Smedstad OM, Meinen CS, Kourafalou V, Hogan P. Temporal variability of the Florida current transport at 27°N. In: Long JA, Wells DS, editors. Ocean circulation and El Nino: new research. New York: Nova Science Publishers; 2009. p. 119–37.

Pinkel R, Smith JA. Repeat-sequence coding for improved precision of doppler sonar and sodar. J Atmos Ocean Technol. 1992;9:149–63. https://doi.org/10.1175/1520-0426(1992)009<0149:rscfip>2.0.co;2.

Pope RM, Fry ES. Absorption spectrum (380–700 nm) of pure water. II. Integrating cavity measurements. Appl Opt. 1997;36(33):8710.

Preston-Thomas H. The international temperature scale of 1990 (ITS-90). Metrologia. 1990;27(1):107.

Sarmiento JL, Gruber N. Ocean biogeochemical dynamics. Princeton: Princeton University Press; 2006. ISBN: 9780691017075. 528 pp.

Schmidt WE, Woodward BT, Millikan KS, Guza RT, Raubenheimer B, Elgar S. A GPS-Tracked Surf Zone Drifter. J Atm Ocean Tech. 2003;20:1070–5.

Seidel MP, De Grandpre MD, Dickson AG. A sensor for in situ indicator-based measurements of seawater pH. Mar Chem. 2008;109:18–28.

Sieracki CK, Sieracki ME, Yentsch CS. An imaging-in-flow system for automated analysis of marine microplankton. Mar Ecol Prog Ser. 1998;168:285–96.

Strickland JDH, Parsons TR. A practical handbook of seawater analysis, Bulletin, vol. 167. Ottawa: Fisheries Research Board of Canada; 1972. 310 pp.

Sverdrup HU, Johnson MW, Fleming RH. The oceans, their physics, chemistry, and general biology. New York: Prentice-Hall; 1942. p. c1942. http://ark.cdlib.org/ark:/13030/kt167nb66r/.

Takeshita Y, Martz TR, Johnson KS, Dickson AG. Characterization of an ion sensitive field effect transistor and chloride ion selective electrodes for pH measurements in seawater. Anal Chem. 2014;86(22):11189–95. https://doi.org/10.1021/ac502631z.

Transportation Research Board and National Research Council. Oil in the sea III: inputs, fates, and effects. Washington, DC: The National Academies Press; 2003. https://doi.org/10.17226/10388.

Werdell PJ, Bailey SW, Franz BA, Harding LW Jr, Feldman GC, McClain CR. Regional and seasonal variability of chlorophyll-a in Chesapeake Bay as observed by SeaWiFS and MODIS-Aqua. Remote Sens Environ. 2009;113:1319–30.

Williams J. Oceanographic instrumentation. Annapolis: United States Naval Institute Press; 1973. 189 pp. ISBN: 0-87021-503–5.

Zeebe RE. History of seawater carbonate chemistry, atmospheric CO_2, and ocean acidification. Annu Rev Earth Planet Sci. 2012;40:141–65. https://doi.org/10.1146/annurev-earth-042711-105521.

Zhou Z, Guo L, Osburn CL. Fluorescence EEMs and PARAFAC techniques in the analysis of petroleum components in the water column. In: McGenity T, Timmis K, Nogales B, editors. Hydrocarbon and lipid microbiology protocols, Springer protocols handbooks. Berlin/Heidelberg: Springer; 2015.

Chapter 3
Platforms for Coastal Ocean Observing

Abstract Instrument platforms, static or mobile, are essential to the observing mission providing not only load bearing surface and housing but also electrical power, telemetry, and, in case of mobile systems, propulsion and navigation. Instrument platforms have evolved from shoreline pilings and manned surface vessels to sophisticated systems including instrumented buoys; autonomous underwater vehicles and other robotic vehicles on, in, or near the water (such as shore-based emplacements); and aircraft or satellite platforms for remote sensing. In this chapter, we discuss both the variety of platforms available and best practices of coupling instrument to platforms.

Keywords Docks · Towers · Pilings · Buoys · Bottom emplacements · Manned vessels · Autonomous vehicles · Satellites

3.1 Fixed Ocean Observing Platforms

3.1.1 Land-Based Ocean Observing Platforms

Tidal observations are undoubtedly the pioneers of ocean observing being mentioned by Aristotle (384–382 BC) and Ptolemy (90–168 AD). Initial quantification was performed using a graduated *tide staff* affixed to an appropriate dock, bulwark or piling. Poor precision caused by sometimes violent wave action led to the development of *stilling wells*, partially enclosed cylinders that smooth out violent short-term surface displacements. Mechanical floating recorders could then provide accurate tidal readings. Stilling wells are essentially vertical pipes mounted across the air-water interface communicated to the outside medium across perforations and/or baffles designed to mechanically dampen wave action in order to allow a more accurate reading. Depth of the well that must at all time traverse the air-sea

The original version of this chapter was revised. A correction to this chapter can be found at https://doi.org/10.1007/978-3-319-78352-9_9

J. E. Corredor, *Coastal Ocean Observing*,
https://doi.org/10.1007/978-3-319-78352-9_3

Fig. 3.1 Stilling well at the Magueyes Island NOAA tide station in southeastern Puerto Rico incorporating a pneumatic bubble gauge

interface depends on magnitude of local tides, where tides are great, stilling wells may be required at lengths greater than 10 m introducing several sources of uncertainty to the reading. Modern pneumatic and acoustic tide sensing systems incorporate the *stilling well* design (Fig. 3.1).

Meteorological systems, measuring such variables as temperature, humidity, wind, atmospheric pressure and solar irradiance, constitute essential units for coastal ocean observing. Instrument platforms as simple as telephone posts, or a variety of dedicated structures in the form of tripods or guywire-supported towers may be emplaced on land or upon existing docks. Lightweight aluminum open frames are favored for tower construction. Such platforms are normally mounted on concrete pads. Commercial housing sheds, analogous to those seen along major highways controlling traffic lights and other systems, can be used to house electronics and computers driving such mentioned systems. These sheds can incorporate air conditioning units and power supplies and can be equipped with satellite and/or cellular communication antennae and global positioning system (GPS) antennae for accurate timing.

In recent years, synchronized high frequency radar transmitter/receiver (Tx/Rx) arrays have become useful for synoptic measurement of surface currents within about 200 km of the coast. Antennae and associated electronics are deployed on land as close to shore as possible or on existing docks. Antennae are mast mounted and masts in turn are mounted on fixed or swiveling bases on concrete pads when possible.

3.1.2 Ocean-Based Ocean Observing Platforms

Floating Bottom-Tethered Buoys

Payloads aboard fixed buoys are integrated into flotation structures of various designs. *Pan* forms (Fig. 3.2) are common, but spheres, monohulls, catamarans, vertical spars, and other designs attend to specialized applications. Buoyancy can be provided by metal or fiberglass hulls or by different polymeric formulations such as syntactic and ionomeric foams. Deep sea floatation units favor glass spheres encased in hard polymer shells.

Mooring materials include chain link, wire rope, synthetic rope, and elastic components. A combination of link chain at the buoy and anchor ends and synthetic wire rope in between minimizes tackle weight. Mooring anchors made up of sets of two or three railroad wheels (weighing about one metric ton each) cojoined through the axes provide a compact modular solution. Other anchor alternatives are scrap chain, properly decontaminated junk motor blocks or purpose-built armored concrete blocks (Fig. 3.3). In coastal shelf moorings, excess chain beyond the tidal datum may be specified such that some chain remains on the bottom even under the highest wave conditions. This *slack chain configuration* is effective in ensuring buoy integrity in heavy seas but results in bottom scouring around the mooring anchor as the varying tides, winds, and currents drive the buoy. Deployment sites devoid of major biota are consequently sought to minimize environmental damage. Loose gravel or sand, subject to recurrent environmental bottom scouring, are preferred. A self-standing frame with lead-filled legs can assist in ballasting the buoy at sea, and facilitates maintenance on land. It can furthermore serve as a mounting surface for instruments intended to monitor near-surface seawater properties. Additional oceanographic instrumentation may be deployed at various depths pertinent to local environmental variability. Such instrumentation is deployed within purpose-built protective cages, integrated to the anchor line using specialized shackles and terminal couplings. Public or commercial satellite systems are used for data telemetry. For nearshore applications, with appropriate antennae, redundant telemetry can be achieved using commercial cellular networks or HF radio. Solar panels and wind turbines can provide system power to the buoys. Power storage is economically achieved using deep-discharge marine lead-acid batteries which double as additional ballast at the bottom of the buoy payload bay.

Fig. 3.2 ODAS buoy in preparation for deployment south of St. John, US Virgin Islands. Acoustic and rotor anemometer and meteorological instruments, in addition to GPS and communications antennae, are emplaced on the top railings and tower struts. An acoustic Doppler current profiler to be mounted integral to the mooring chain is enclosed in a stainless steel cage alongside the buoy base. A conductivity temperature/instrument is mounted horizontally on the right-hand base strut. The cylindrical instrument and battery well protrudes above and below the pan-shaped yellow buoyancy foam girdle. Solar photovoltaic panels provide electrical power

Coastal ocean observing buoys classified as ocean data acquisition systems (ODAS) are multipurpose buoys bearing meteorological and oceanographic instrumentation. The main buoy components are the anchorage and buoyancy systems and the payload (incorporating power system, sensors, data processing and storage systems, and the telemetry system). Additionally, safety provisions include lights,

Fig. 3.3 Wave buoy and bottom mooring. Buoy to left, concrete anchor and chain to right

radar reflectors, lightning rods, and even bells or foghorns in high traffic areas subject to frequent fogs. GPS capability alerts to loss of *station keeping* and allows tracking of buoys accidentally set adrift. Buoy software is programmed to send emails or place cellular telephone calls if the GPS signal is outside the buoys preestablished watch radius.

Buoys are equipped with watertight payload bays housing batteries and electronics and communications equipment. Instrument towers or masts bear meteorological sensors, antennae, foghorns, radar reflectors, lighting rods, and other instruments.

NOAA's National Data Buoy Center (NDBC) operates a wide variety of data buoys ranging from small 1.5 diameter COLOS buoys for inshore waters to the large 12 m diameter offshore DISCUS buoys. While primarily meteorological, NDBC buoys usually bear wave-sensing accelerometers and instruments for measuring surface seawater temperature and salinity. Deep water ODAS moorings incorporate several additional features such as extended lengths of buoyant polypropylene rope along the mooring line. Judiciously positioned subsurface buoys and lead sinkers achieve additional resilience to wave action through the formation of running-S inverse catenary forms.

Incorporating most elements described, the Gulf of Maine Ocean Observing System (GoMOOS) ODAS moored buoys (Wallinga et al. 2003) have been successfully deployed in the Gulf of Maine, the Yucatan Channel, and Caribbean and Atlantic waters of Puerto Rico and the Virgin Islands. These buoys are routinely recovered, refurbished, and redeployed, and instruments are replaced with freshly recalibrated units (Fig. 3.2).

Fig. 3.4 Buoy mounted carbon dioxide measurement system. Left: detail of the mooring system with a chain mooring on the left and a pre-tensioned three-member elastic mooring on the right. Right: newly refurbished MAPCO2 buoy equipped with underwater dye-based pH sensor and CTD in addition to onboard pCO_2 sensors

Specialized buoys may come in different forms. Stainless steel, aluminum, or fiberglass hulls provide the required buoyancy. Moorings can be simple single-line constructs or complex systems incorporating dual anchors, elastic rubber bands, and S-shaped inverse catenary forms incorporated into the nonelastic anchor line to dampen chaotic buoy movement. Small spherical vessel designs under 1 m diameter are favored for dedicated wave measurement buoys to maximize buoy response to wave forcing (Fig. 3.3).

The MapCO2 buoy, designed and operated worldwide by the NOAA Pacific Marine Environmental Laboratory (PMEL) Carbon Program, contains an onboard CO_2 equilibrator for measurement of atmospheric and seawater partial pressures. Submerged mounting sites are available for CTD and additional carbon chemistry instrumentation. This buoy features a polyurethane foam-filled fiberglass hull. For safe deployment on coral reefs along the high wave action fore-reefs, the buoy incorporates a two-point mooring with a taught triple-member elastic band on the seaward end (Fig. 3.4).

Sea Bottom Emplacements

Stationary sea bottom emplacements (SBEs) are commonly used to mount *upward looking* acoustic Doppler current profilers (ADCP) for operational current profiling throughout the water column to depths of about 1 km. Simple SBE instrument

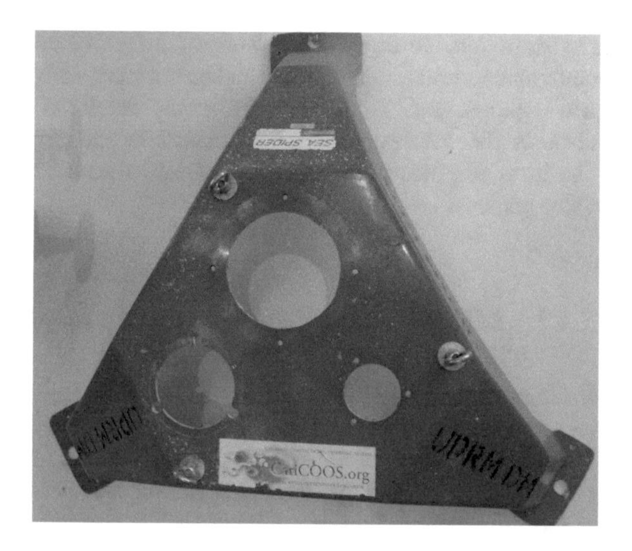

Fig. 3.5 Trawl resistant frame for instrument emplacement on the ocean bottom

mounts built of aluminum or fiberglass can incorporate disposable ballast for ease of recovery, either separate or as part of the base. To avoid entanglement in bottom trawls, these stationary sea bottom emplacements can incorporate *trawl resistant* features. Cages with smooth sloping sides devoid of external protrusions allow ballasted bottom fishing trawls to slide over the housing minimizing damage (Fig. 3.5).

More elaborate SBE bear vertical profilers that periodically sample the water column from the base to (or near to) the ocean surface using instrument packages incorporating a wide variety of sensors for physical, chemical, and biological observations. Some designs feature bottom mounted winches that periodically deploy and recover buoyed instrument packages. A different approach to buoyed spooled profilers incorporates *crawler* vehicles that travel along taught lines from bottom mounts to near-surface buoys and back.

Further technological developments include multiple SBE linked by sea-bottom cabled and acoustic communications and power nodes. Research observatories currently in use and under implementation incorporate sophisticated *node junction boxes* that allow mating of diverse instruments to the network. Instrumented autonomous vehicle fleets, using node-based acoustic beacons, navigate precise trajectories to add further data layers to observation-based 4-D documentation of ocean properties.

3.2 Mobile Ocean Observing Platforms

Manned, mobile, ocean observing platforms include ships and submarines and their associated vertically deployed sensor packages, towed fixed depth or undulating vehicles, or tethered remotely operated vehicles. Manned aircraft and space stations serve as remote sensing platforms. Unmanned mobile on-water or in-water platforms include free-floating buoys (surface or profiling), autonomous surface craft, autonomous underwater vehicles, buoyancy-driven gliders, and the innovative wave gliders and saildrones. Unmanned satellites now constitute an essential resource for remote sensing of ocean surface and near-surface properties. All such platforms can incorporate a variety of instruments and sensors including still and video cameras.

3.2.1 Manned Vessels and Shipboard Deployed Vehicles

Instruments deployed on lowered cables have been a fixture of oceanographic research since the HMS CHALLENGER expedition (1872–1976) when thermometers and sampling bottles were routinely deployed to full ocean depth using hemp rope and lead weights, including, remarkably, the sounding in the Marianas Trench of the aptly named *Challenger Deep*, the deepest part of the world ocean at greater than 11 km. Today, the carrousel or rosette sampler (a cylindrical frame fitted with oceanographic sampling *bottles*, sensors, and data links) is a common sight aboard oceanographic and survey vessels.

Hydraulic or electric spooling winches equipped with single- or multiple-conductor electromechanical cable deploy these carrousel-like platforms. *Slip-ring* devices mounted on the winch axle allow instrument data and ship-board commands and power for bottle operation to flow through the cable's electrical conductors to and from the operation consoles so that instrumental data output may be viewed in real time and bottles may be remotely triggered from the surface vessel at standard or selected depths (Fig. 3.6).

These instruments are usually deployed from a stationary vessel which contributes to the expense of their operation since ship time can be valued in the tens of thousands of dollars per day and a shallow deployment can consume 1 h or more of ship time. To address this issue, fast CTDs coupled to specialized underway winches now allow profiles to depths of 400 m while underway at 10 knots but cannot retrieve subsurface water samples.

A different approach to these recoverable instruments is embodied in expendable instruments such as the expendable bathythermograph (XBT), a probe developed originally for defense applications. This instrument consists of an expendable ballasted probe carrying a thermistor sensor connected through a launcher to a shipboard console. Upon descent, the streamlined probes and the handheld or deck-mounted launcher both unwind thin conductive copper wires from internal spools. The probes rapidly achieve terminal velocity allowing calculation of depth from the time the probe strikes the water dispensing with the need for a separate pressure (depth) sensor. Expendable CTDs, based upon this model, but with an additional conductivity sensor, have been available for some time. Such probes can be deployed from a moving vessel.

Modern research and survey vessels also mount flow-through instruments that allow continuous sampling of near surface waters while underway. Water is taken in

Fig. 3.6 Twenty-four bottle carrousel fitted with 10 l bottles. A CTD equipped with chlorophyll, dissolved oxygen, and turbidity sensors is mounted in a stainless steel cage below the bottles. *Red bricks* are lead weights to facilitate deployment. The operating *winch* equipped with electromechanical cable is to the right of the carrousel

across a dedicated through-hull sampling port or from the vessel sea chest and is served through a vortex *debubbler* mechanism since most flow-through instruments are sensitive to bubbles in the stream. Instrument data is time-stamped and georeferenced using the ship's navigation data. For gas exchange studies, bow-mounted air samplers feed near-surface air to the shipboard laboratory. Extreme care must be taken to avoid contamination from the ship's exhaust.

Towed instruments achieve measurements that would otherwise be impossible directly from the moving vessel. Specialized instruments may have a custom-built vehicle but various commercial tow vehicles are also available. Towed vehicles, purpose-built for single instruments, are commonly enclosed within a torpedo shaped housing to minimize drag. Some may need to be towed close to the bottom such as magnetometers that detect sub-bottom magnetic anomalies or side-scan sonar devices that achieve 3D ocean imagery using high-frequency but short-range acoustic pulses.

Other towed vehicles can achieve undulating patterns descending to depths exceeding 100 m and ascending to near the ocean surface. Some undulating vehicles take advantage of the ship movement to power a propeller-driven onboard dynamo. This dynamo in turn powers a servo-operated elevator wing guided by the onboard pressure sensor to achieve the undulating pattern within a preset depth range. Winches used to deploy these platforms are equipped with conductor cable and slip-rings for real-time vehicle control and data retrieval (Fig. 3.7).

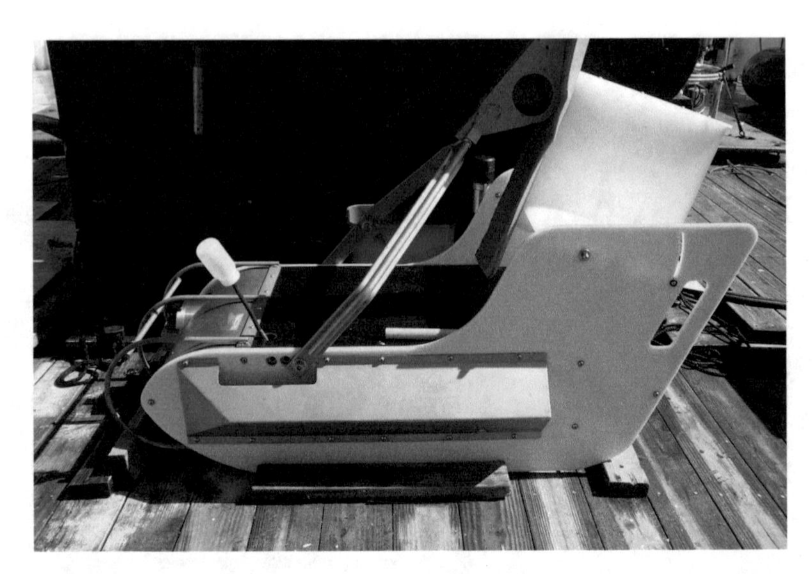

Fig. 3.7 Towed undulating vehicle. The vehicle pictured is fitted with CTD, chlorophyll fluorometer, and dissolved oxygen optode

Although tethered, remotely operated vehicles (ROV) can be deployed from diverse platforms, including docks and bridges, shipboard operation is common. These vehicles are attached to the vessel or shore-side control system with an *umbilical* cable carrying power and data. Some ROVs operate on the ocean bottom rolling on tracks but most are buoyant and carry three-axis propellers for navigation in the water column. They are commonly equipped with video and audio, remote manipulators, and sample holding baskets. Shipboard or shore-based operators navigate and manipulate samples or equipment remotely using the video feed. Some ROVs are dedicated to coastal research, exploration or archeological research, but most are used in industry for equipment inspection, maintenance, and repair.

An alternative to high cost research and survey vessels is that of *ships of opportunity*; cargo ships, ferries, cruise ships, or other commercial vessels traveling recurrent routes. Such manned vessels may be fitted with automated XBT launchers for vertical temperature soundings and/or autonomous flow-through instrument suites for monitoring near-surface conditions. The European *FerryBox* system, a standardized instrument package for operation aboard ships of opportunity, has proved to be quite successful. These systems are fed from the ships sea chest or dedicated through-hull water intakes and are fitted with de-bubblers to remove entrained air bubbles from the seawater stream and filters to remove coarse sand and other entrained mineral particulates. Analytical flow-through systems commonly include T/S devices, chlorophyll fluorometers, oxygen optodes or polarographs and optical turbidity sensors. If fitted with nutrient sensors, additional hollow fiber cross-flow filters may be required to remove plankton and other finer particulates prior to analysis. The systems require periodic maintenance to remove debris and combat biofilm buildup (see Sect. 4.4).

3.2.2 Lagrangian Drifters

Drifting buoys devoid of instrumentation constituted the basis for Lagrangian current measurements for many years before the advent of modern electronics. *Current slip* (decoupling of buoy movement from current drift) is minimized using submerged ballasted *drogues* tethered to the buoy at the depth of interest. Drogues were often constructed with sheet metal or plywood cross-planes but even aviation parachutes were successfully used. Curiously, piano wire was the tether of choice for deep water deployments, practicable but laborious, to depths down to 1 km. Drogue buoys were followed at sea and vessel navigation with the buoy alongside was used to plot the buoy trajectory.

Visual location and triangulation using a compass and sextant or transit was a convenient method for obtaining coastal current data by deploying appropriately drogue-stabilized buoys and carrying out repeated simultaneous readings from two or more land-based stations of known location, often capes or promontories. This effort however could only be maintained over periods of hours and nocturnal readings required a light beacon aboard the buoy.

Two global drifter programs contribute to advancing coastal ocean observations through their use in calibration of satellite temperature data thus providing sea surface temperature maps extending into the coastal zone. The Surface Velocity Program, an offshoot of the World Ocean Circulation Experiment (WOCE), arose from the need to accurately determine surface drift in the world oceans. Several United Nations agencies as well as private and state institutions from member nations participated during WOCE and continue to do so in the Global Drifter Program now a part of the Global Ocean Observing System (GOOS). The stated objective of the GDP is:

> *To meet the need for an accurate and globally dense set of* in-situ *observations of mixed layer currents, sea surface temperature, atmospheric pressure, winds and salinity.*

As implied, GDP buoys are currently equipped with temperature, atmospheric pressure, and conductivity sensors in addition to the original telemetry/navigation instrument suite. Rather than using the GPS system, these platforms rely on the Doppler shit of a buoy signal received by the *ARGOS* satellite constellation allowing precision of between 150 and 1000 m as required. Central to the program is the Holey Sock Buoy (*holey* because of the holes in the unusual drogue design consisting of a wide diameter hoop-reinforced synthetic fabric cylinder with large holes in the fabric). This design has been shown to minimize slip at an effective depth of 15 m to 0.7 cm/s in 10 m/s wind speed. For perspective, it is noted that a comparable buoy that has lost its drogue would slip downwind at 8.6 cm/s under such winds. The buoys are designed for extended endurance at sea. At this writing 1418 are operationally deployed throughout the world ocean.

Part of the UN-sponsored global GOOS program and of the Global Ocean Climate Observing System as well, the advanced *ARGO* profiling buoy system provides data to depths of 2000 m on temperature, salinity, and, in later versions, dissolved oxygen. Injecting buoyant fluids from within a cylindrical pressure hull into external bladders varies their effective buoyancy allowing repeated ascents and descents. To conserve power and avoid biological fouling in the sunlit near surface ocean layers, the buoys keep a *parking depth* of about 2000 m and ascend at approximately 10 day intervals. Following satellite positioning, the buoy transmits its data package and descends to the parking depth. The ARGO system is a major success for the global oceanic and atmospheric climatology community. Originally developed as the *Autonomous Lagrangian Circulation Explorer* (ALACE) (Davis et al. 1992), as of February 2017, 3962 ARGO floats are in operation throughout the world ocean.

Modern *coastal* Lagrangian drifting buoys are equipped with GPS, and telemetry capabilities using communication satellites or commercial land-based cellular networks. A widely used commercial system incorporates a small spherical buoy tethered to a diamond-shaped drogue constructed with a plastic tubing frame and a synthetic fabric drogue. Electronics, including GPS, optional temperature reading, a data logger, and satellite communications equipment, are housed within the buoy. Battery life allows about 1 week operation. Larger commercial systems for more extended deployment incorporate the holey sock design.

Low cost drifters incorporating consumer market GPS navigation and VHF radio transmitter units can be built with readily available construction materials such as closed-cell foams and PVC pipe. Such systems are useful for shorter term deployments with higher data return since they can be deployed and recovered from small boats. VHF signal reception is however strictly in line-of-sight from the transmitter so effective range from small boats is limited to a few kilometers for these applications.

3.2.3 Autonomous Surface Vehicles

Satellite global navigation systems together with global telemetry now allow geolocation of remote surface platforms and two-way communication with such vehicles. Solar, wind, and wave power in turn allow for extended vehicle autonomy. Several innovative *autonomous surface vehicles* (ASV) are now available on the market, a few being civilian offshoots of naval *target drones*. These AUVs can be operated remotely but can also cover waypoint grids autonomously. Vessels can be single-hulled but the twin-hulled catamaran rig is favored for stability. The US National Oceanic and Atmospheric Administration (NOAA) currently operates small mono-hull ASVs for inshore bathymetric surveys. Many have also found operational application in inshore surveillance of ports and harbors and other high value assets. Surveillance vehicles can be fitted with diesel engines but autonomy is poor. In general, long-endurance vehicles make use of solar, wave, and/or wind power.

Among the most innovative and a potential harbinger of change is a wave-powered surface platform that has gained widespread recognition. Featuring a surfboard-like low-profile single hull (3 m LOA), this *unmanned surface vehicle* (USV) is propelled by an array of six movable winglet pairs mounted on a subsurface frame. Vehicle and propulsion system are connected by a 6–8 m tether. The difference between heightened orbital wave-induced motion at the surface and minimal movement at depth coupled to the movable winglets provide propulsion power for speeds up to 3 knots. Payload power is provided by a solar panel array supplying a lithium-ion battery bank. A directional thruster mounted aft like a conventional ship's propeller but capable of swiveling on the horizontal plane allows surface navigation.

These USVs are capable of year-long deployments and can carry a large number and variety of instruments, onboard, subsurface, or towed. Patrolling in single units or *swarms* or holding station alone or in arrays, these USVs achieve significant cost reduction for surface and subsurface observations. Recurrent operational coastal IOOS applications are still lacking, but the US Pacific Islands Ocean Observing System (PacIOOS) has carried out experimental missions in the vicinity of the Hawaiian Islands recording wave height, direction, and period and, notably, traversing and surviving 7 m hurricane wavefields. Custom towed bodies have found applications in geophysics, acoustic fisheries surveys, and ocean optics. Successful fish stock assessment missions using towed sonar arrays have been reported off South Africa. Lately a NASA wave powered surface vehicle for satellite calibration equipped with an onboard optics package, a towed submersible body with multispectral sensors and redundant telemetry antennae has been deployed successfully in coastal waters off the islands of Hawaii and Puerto Rico (Fig. 3.8).

A versatile *sailing* autonomous surface vehicle is now also being deployed for ocean observing. The vehicle features a 7 m outrigger-equipped hull and a rigid sail structure extending to a height under 5 m. Manufacturers specify endurance up to 1 year, wind-dependent speed of up to 8 knots and payload capacity of up to 100 kg.

Fig. 3.8 Wave powered surface vehicle. The vehicle pictured is equipped with an optical array for vicarious calibration of optical satellite sensors. The faired yellow cable upon the vehicle deck is used to tow a submersible optical package

The platform can be fitted with a variety of atmospheric and oceanographic instrumentation. In July 2017, the Alaska Ocean Acidification Network, coordinated by the Alaska Ocean Observing System, launched a pair of these vehicles equipped with an advanced version of the MAPCO2 instrument (ASVCO2) with the mission of characterizing ocean acidification in remote Bering Sea and Arctic waters.

3.2.4 Underwater Gliders

The technology of underwater gliders, developed initially by Douglas Webb, building upon the successful developments of profiling floats, was first proposed and popularized in a visionary publication by Henry Stommel (1989) describing a new profiling float with wings that allowed horizontal displacement during dive and ascent, a vehicle that he named the Slocum float in honor of Joshua Slocum, the first single-handed world circumnavigator. Stommel envisioned a worldwide fleet of these profilers operated by mission control of a World Ocean Observing System. Stommel saw the grand opportunity to garner a harvest of data on a global deep-ocean basis far in excess of that available with, by then traditional, CTD casts from ocean-going vessels. Webb went on to build and commercialize the first operational gliders.

Today the operational fleet of gliders exceeds 1000 units used for multiple applications including defense. Widely available commercial vehicles are sufficiently small to be deployed and recovered from small vessels (even rubber dinghies) by two people. Hulls, usually less than 2.5 m length overall, are roughly torpedo or tear-drop shaped with cylindrical hulls, bow bulbs, and faired sterns, or may exhibit more hydrodynamic spindle-formed external fairing covering the pressure hull. Most models feature wings to achieve hydrodynamic lift with spans typically under about 120 cm (Fig. 3.9). A European product dispenses with the wings achieving lift through hull design alone.

In contrast to the profile achieved by the ARGOS float dependent on current drift, the wings allow the glider to achieve a slanted glide path resulting in a saw-tooth pattern of descents to depths to 1500 m or shallower as required, and subsequent ascent. Profiling gliders have most certainly transformed the field of ocean observing increasing the spatial and temporal data coverage from the world ocean and coastal seas (Rudnick et al. 2004).

Vehicle pitch and thus the descent angle are controlled in most gliders by shifting the battery pack internally fore or aft. Steering on the horizontal plane is achieved either by using a stern mounted rudder much like an airplane or by an ingenious system that moves the battery package laterally along a semicircular rack and pinion arrangement inside the pressure hull thus shifting the center of mass laterally. Redundant stern- or wing-mounted antennae provide satellite communication by various means for positioning data relay and vehicle control. Upon surfacing, the antennae pierce the surface allowing radio communications. Payload bays

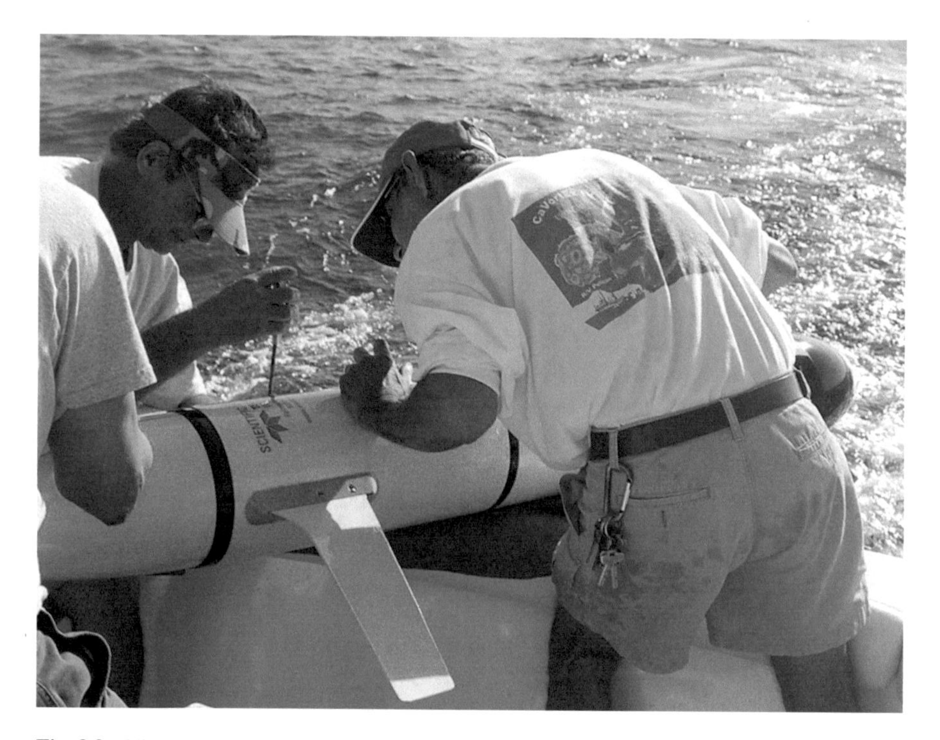

Fig. 3.9 Glider recovery operation. The glider pictured is equipped with CTD, inertial navigation system, GPS, and satellite antennae. Wings are being removed for safe handling

accommodate a wide variety of physical, chemical, optical, and acoustic sensors, many custom designed for gliders (Rudnick and Perry 2003). Gliders are relatively slow, achieving forward speed of 0.2–0.4 m.s^{-1} (about 0.4–0.7 km.h^{-1}) in the absence of contrary currents. Oceanographic cross sections thus obtained are consequently not truly synoptic and transient features may be blurred by Doppler smearing (Rudnick and Cole 2011).

Similar to the ARGO floats discussed above, buoyancy control on gliders is achieved by pumping a buoyant fluid between an internal reservoir and an external bladder. Electrical pumps driven by onboard disposable alkaline or reusable lithium batteries are most common. The latter typically provide up to 17.5 mega joules (MJ) of available power and allow missions extending to as long as 10 months with ranges to 10,000 km. An alternative energy source for offshore applications is found in the *thermal glider* approach using a phase-changing wax that contracts and expands as it transitions from liquid to solid as the vehicle dives from the warm surface water into colder deeper water. Applying and releasing the pressure thus generated to compress a gaseous nitrogen reservoir provides the power to pump the working fluid and reestablish buoyancy once the target depth is reached. Such a scheme is limited only by mechanical failure, but again, is more suited for deep diving large area surveys than to nearshore coastal operations.

Today, gliders are used primarily for research purposes but many of these field campaigns are coordinated and jointly supported by ocean observing systems that archive and display data and assimilate data into regional forecast models. Glider lines are maintained by the CALCOFI survey discussed above, initially based on manned oceangoing vessels and now supplemented by glider lines along various preestablished survey lines normal to the west coast of the USA. The University Center for Ocean Observation Leadership (RUCOOL), formerly the Rutgers University Coastal Ocean Observation Laboratory, operates periodic cross-shelf transects on the wide eastern shelf of the US mid-Atlantic coast and is recognized for having achieved the first transatlantic glider crossing remotely navigating a Slocum glider from the US coast to the coast of Spain over a period of 221 days in 2009. The Caribbean Coastal Ocean Observing System (CariCOOS), a component of the US IOOS program, in collaboration with NOAA's Atlantic and Meteorological Laboratory (AOML), seasonally operates a pair of gliders, one in Atlantic Ocean waters off the north coast of Puerto Rico and one in Caribbean waters off the south coast of Puerto Rico for hurricane monitoring and research.

3.3 Ocean Observing Satellites in Terrestrial Orbit

Satellite oceanography is a mature field using a number of platform types, sensor payloads, and technologies (Robinson 2010). Many satellites serve not only as sensor platforms but also as data relay platforms receiving data from land-/ocean-based autonomous sensor systems and relaying the data to the operator. Additionally, specific satellite constellations are dedicated to navigation and positioning providing services essential to real-time ocean observing. Sensor systems in turn can be divided into those bearing passive sensors alone which receive reflected radiation and active systems that incorporate a radiation source and a backscatter receiver.

Operational satellites travel either in the uppermost range of low earth orbit, at altitudes of between 700 and 1500 km or in geostationary orbit at an altitude of 35,786 km. Operational environmental satellites in low earth orbit include various US, EU, and Asian weather and research satellites or satellite constellations that bear atmospheric terrestrial and ocean sensors as well as data relay systems. Environmental satellites in low earth orbit are mostly sun synchronous, and in near polar orbit at altitudes between 700 and 1350 km. Slight inclination of the orbit with reference to the polar axis of rotation allows *orbital precession* such that the angle of solar illumination is invariable except for the apparent seasonal solar displacement. These orbits allow daylight images and near total global coverage with 3–35 orbit repeat periods. For scanning sensors, swath widths of a least 2700 km allow daily coverage.

Geostationary orbit at an altitude of 35,786 km above the equator allows a satellite to orbit at the same angular rotation as the earth, thus appearing to remain fixed above a particular point of the equator. The greater distance from the earth results in telecommunications delays of about 2 s hampering real-time communications protocols but useful for data relay. Radiometers aboard geostationary weather satellites such as the US GOES constellation allow continuous mapping of sea surface temperature (SST) at 4 km for the planetary disks in view.

Standard sensor systems aboard US and European operational environmental (weather) satellites include the Advanced Very High Resolution Radiometer (AVHRR), providing high-resolution visible, near, and far or thermal IR imagery, the latter allowing sea surface temperature (SST) retrieval. The European Union's ESA European environmental ENVISAT satellite mission (2002–2012) carried a multispectral visible/ IR scanner (MERIS) together with a wide array of land, ice, and atmosphere sensors. NASA Earth Observing Mission satellites Aqua and Terra carry ocean color sensors as well as a variety of other sensors. Currently, operational MODIS instruments measure ocean color in nine bands in the Visible/IR across a swath width of 2300 km at 1 km resolution. VIIRS, a commercial instrument aboard a NOAA weather satellite, similarly uses nine ocean color bands in the Visible/IR range over a swath width of 3000 km and a Horizontal Interval on Ground (i.e., pixel resolution) <1.6 km at the end of the scan, hence considerably better at nadir. Satellite-borne sea surface temperature (SST) radiometers, ocean color sensors, and sea surface salinity sensors are discussed in greater detail in Chap. 2.

Active instruments aboard dedicated satellites provide imagery ranging from the fine-scale capability of detecting ship wakes using high-resolution synthetic aperture radar to the detection of sea surface height anomalies on the order of tens of centimeters due to ocean current gradients and ocean eddies across horizontal ranges of tens to thousands of kilometers. The Ocean Surface Topography Mission, a joint effort of NASA, NOAA, CNES, and EUMETSAT, currently operates the Jason-2 satellite in continuation of the TOPEX/POSEIDON and Jason-1 missions. This satellite transmits microwave pulses to the ocean surface and captures backscattered radiation. Precise timing and atmospheric corrections allow sea surface height calculation to a precision of about 5 cm. Such data makes its way into operational coastal ocean observing since it is incorporated into global circulation models which in turn provide boundary conditions for nested regional and subregional models.

In addition to the primary role of GPS in positioning and navigation, the high accuracy of the atomic clocks aboard these satellites provides temporal synchronization for a wide range of applications ranging from cellular communication networks to electrical power grids. In ocean observing, land- or buoy-based HF antenna radar arrays for surface current measurement use this timing for synchronization between stations to alternate *transmit* (Tx) and *receive* (Rx) functions between stations (Lipa et al. 2009), an application termed *bistatic* operation.

References

Davis RE, Webb DC, Regier LA, Dufour J. The autonomous Lagrangian circulation explorer (ALACE). J Atmos Ocean Tech. 1992;9:264–85.

Lipa B, Whelan C, Rector B, Nyden B. F Radar bistatic measurement of surface current velocities: drifter comparisons and radar consistency checks. Remote Sensing. 2009;1:1190–211. https://doi.org/10.3390/rs1041190.

Robinson IS. Discovering the ocean from space: the unique applications of satellite oceanography. Berlin/Heidelberg: Springer; 2010. 638 p. ISBN 978-3-540-68322-3.
Rudnick DL, Perry MJ, editors. ALPS: autonomous and Lagrangian platforms and sensors, workshop report. 2003. 64 pp. www.geo-prose.com/ALPS.
Rudnick DL, Cole ST. On sampling the ocean using underwater gliders. J Geophys Res. 2011;116:C08010. https://doi.org/10.1029/2010JC006849.
Rudnick DL, Davis RE, Eriksen CC, Fratantoni DM, Perry MJ. Underwater gliders for ocean research. Mar Technol Soc J. 2004;38:73–84. https://doi.org/10.4031/002533204787522703.
Stommel H. The Slocum mission. Oceanography. 1989;2:22–5.
Wallinga JP, Pettigrew NR, Irish JD. The GoMOOS moored buoy design. OCEANS 2003. Conf Proc. 2003;5. https://doi.org/10.1109/OCEANS.2003.178318. http://oceanobservatories.org/site/ce01issm/.

Chapter 4
Environmental Constraints to Instrumental Ocean Observing: Power Sources, Hydrostatic Pressure, Metal Corrosion, Biofouling, and Mechanical Abrasion

Abstract Electronic instruments and platforms, especially the mobile platforms, require a stable source of electrical power which in most cases is not available from a large-scale terrestrial power grid. Batteries alone for submerged applications or combinations of solar cells and batteries for surface applications are the principal power sources now in use, but wave, wind, and other power sources are becoming available. Significant power economy afforded by modern electronic systems now extends power autonomy beyond 1 year for certain applications. Sensitive electronic components for ocean observing are contained in pressure-resistant housings constructed from corrosion-resistant materials or otherwise protected from corrosion. Pressure housings must be provided with waterproof ports for external sensors, power, and data cables. Metal structures are subject to electrochemical seawater-induced corrosion, in many cases mediated by bacterial metabolism. Metal structures immersed in seawater can be protected using both passive and active strategies, the former employing sacrificial metals and the latter the use of impressed electrical current. Biofouling can severely impair instrument response and vehicle performance. Mechanical means (wipers, shutters, slick surfaces) and/or biocidal means (organometallic formulations) allow reduction or amelioration of biological fouling. Special consideration is due to anchoring systems and electrical wires and cables subject to abrasion in high energy marine environments.

Keywords Pressure housings · Metal corrosion · Biofouling · Mechanical abrasion

4.1 Power Supplies for Autonomous Coastal Ocean Observing Platforms and Instrument Payloads

Unless connected to a land-based power grid, coastal ocean observing platforms must be provided with one or more sources of electrical power for instrument operation, telecommunications, satellite positioning, and, in some cases, climate control. Mobile platforms moreover require propulsive power.

The original version of this chapter was revised. A correction to this chapter can be found at https://doi.org/10.1007/978-3-319-78352-9_9

J. E. Corredor, *Coastal Ocean Observing*,
https://doi.org/10.1007/978-3-319-78352-9_4

Research and survey vessels are, for the most part, equipped with diesel engines and generators providing ample power for all the above requirements. Surface surveillance drones such as those used for port security may also be equipped with diesel engines. Coastal ODAS buoys generally rely on solar panels and conventional marine deep-discharge lead-acid battery banks for house power. Autonomy of buoys so equipped, limited only by battery life and solar panel integrity, extends well beyond the customary 1 year deployment. Wind turbines may be used to supplement the primary solar system. Ocean gliders rely on battery banks using either rechargeable lithium cells or disposable alkaline cells to provide power for pumping the operating fluid and for house power (Rudnick et al. 2004). Sailing drones depend on a combination of propulsive wind power and solar power for payload operation.

High frequency radar systems commonly rely on land-based electrical power grids. However, an autonomous Remote Power Module (RPM) to support extended autonomous HF radar unit operations has been designed, fabricated, and tested in the challenging remote environments of the Arctic and subarctic Alaska (Statsewich et al. 2011). The RPM incorporates three complimentary power sources: solar panels, wind turbines, and a biodiesel power generator, all backed by a battery bank. Additional subsystems include command and control, satellite communications, and power performance monitoring modules. Packaged in small units that two people can carry, the RPM is deployable using small vehicles, boats, or sled-equipped snow machines. Commercialization of such modules will facilitate extension of current networks by allowing grid designs that minimize the number of stations necessary for efficient area coverage. Development of analogous modules incorporating climate control for deployment in tropical regions will further extend possible coverage.

A number of innovative solutions to power requirements of autonomous systems are now in use for industrial applications or are currently under development for low power applications. Dewan et al. (2014) describe electrolytic systems based on sacrificial anode sources, sediment microbial fuel cells based on redox differences between the water column and organic-rich anoxic sediments, and piezoelectric *energy harvesters*. *Thermoelectric energy generators* (solid-state thermal gradient devices) mounted under solar cells are also being put into use to take advantage of the large heat loss of conventional solar cells.

Wind- and wave-powered kinetic vehicles are described above. Kinetic generators using wave and current energy are now commercially available. Such systems are designed for large power output but have yet to be miniaturized for instrument operation (Dewan et al. 2014).

4.2 Ocean Observing Instrument Mounts and Housings

Instruments deployed in the ocean environment are subject to direct seawater exposure under partial or continuous submergence, to great depths in many instances. Many sensors require direct contact with the ocean medium and their electronic circuitry and power sources must be protected from the corrosive properties of seawater and biofouling by aquatic organisms. Moreover, submersible vehicles or submerged emplacements may not provide protection from the environment. Indeed, in many vehicles the payload bays are open to the seawater medium under local hydrostatic pressure. Most submersible instruments are designed as self-contained units encased in their own pressure hull.

Modern instruments designed for submerged autonomous deployment are commonly contained in cylindrical pressure housings or, in some special cases, in spherical housings. Depending on the depth of the application, different pressure hull materials may be used ranging from PVC and acrylic for shallow water applications, to anodized aluminum or DELRIN (a polyoxymethylene polymer formulation of exceptional density, stability hardness, and rigidity) for intermediate depths and stainless steel, titanium, or spherical glass for very deep water.

One or both endcaps on cylindrical housings may be removable for access to electronics and batteries. Such endcaps may be disc-shaped to cover the end of the housing with an O-ring seal flush to the cylinder wall edge. Endcap to cylinder wall junctions may be secured by bolts through the endcap penetrating the threaded cylinder wall or endcap and cylinder may be affixed with nuts and bolts if the cylinder ends are flanged, as is often the case for steel or aluminum housings. More commonly a *shoulder* endcap is machined so that a flange covers the cylinder end and the endcap body penetrates a few centimeters into the cylinder and seals with one or more additional O-rings between endcap and cylinder. Nylon bolts may replace steel bolts to circumvent corrosion but due consideration must be given to their relative fragility.

Specialty steel or titanium cages may be provided for high exposure applications where the fragile instruments may be in danger of severe shock such as vertically deployed instruments or instruments mounted in-line on a mooring cable for which deployment and recovery maneuvers are particularly hazardous. To affix cylindrical pressure housing instruments to the platform deck or the cage structure as may be the case, half-moon indents are machined into PVC or DELRIN base block pairs to fit the cylinder diameter. The instrument is seized to the blocks using various steel or plastic hose-clamp systems or with bolted-on complementary half-moon bracket clamps of the same block material. The base blocks are threaded or perforated for mounting aboard the platform with hex-head bolts or U-bolts. Bottle mounts, common to CTD/carousel packages, may be adapted for instrument deployment aboard the carousel in place of oceanographic sampling bottles.

External sensors such as a thermistors or conductivity bridges require connection across the pressure hull to the power source and system electronics. In addition, data communication ports are usually required. These connections are commonly achieved using waterproof bulkhead-mounted wet-pluggable connectors readily available from specialized manufacturers in a variety of configurations.

Optical sensors may be housed within the pressure housing and need only optical windows which can be integrated into the pressure housing such as is found in fluorometers or optodes. Specialized optical instruments operating in the UV must employ materials transparent to UV such as crystalline quartz for optical windows.

Payload capacity and power limitations common to small mobile platforms such as profiling floats, buoyancy gliders, AUVs, wave gliders, and other lightweight mobile platforms have driven the development of low power, low weight, compact, flush-mounted sensors that can be incorporated into the hull without compromising hydrodynamic vessel properties. Doppler velocity and Doppler navigation loggers, ADCPs, optical instruments such as nephelometers and fluorometers, and a variety of other instruments configured for such applications are now widely available.

4.3 Metal Corrosion Considerations Pertinent to Ocean Observing

4.3.1 Fundamentals of Electrochemistry

Metal corrosion depends on the susceptibility of individual metals to oxidation, susceptibilities that have been well established through experimental assessment. To such an end, all redox half cells, including those where metals participate in the reaction, are compared to a standard *half cell*, the hydrogen reference electrode consisting of a platinum-black electrode immersed in a 1 N HCl solution and bathed in a stream of hydrogen gas bubbles. This electrode is defined, by convention to have a standard half-cell potential of 0 V at standard temperature and pressure (STP: 1 atm pressure and 273.16° K). Electrical potentials established between this absolute reference cell and all other half cells at standard conditions (STP as above and electrolyte concentrations at 1 M) are referred to as standard half-cell potentials. Regardless of the sense of the reaction, equations are written as reductions. The equation for the standard hydrogen half-cell reduction is

$$2H^{+} + 2e^{-} = H_2 \qquad E^0 = 0 \text{ V} \tag{4.1}$$

If the electrode in question undergoes corrosion when in contact with the reference half cell, the resulting potential will be negative, the hydrogen half cell will constitute the cathode, the target half cell will operate as the anode. On the contrary, if the hydrogen gas is oxidized, the potential of the test half cell will be of a positive sign. Potentials of a wide range of half cells have been experimentally determined and are tabulated for ease of use as the electrochemical series (Vanýsek 2010). Standard potentials of the electrochemical series (E^0) range from a low of −3.04 V for the highly corrodible metal lithium to 2.87 V for the highly corrosive fluorine gas, a surprisingly low span of less than 6 V. Platinum (often used as a cathode due to its comparable inertness) exhibits a standard reduction potential of 1.18 V while gold exhibits a potential of 1.692 V. Zinc, in keeping with its easily corrodible nature, exhibits a correspondingly low standard reduction potential (−0.762 V). Potentials of half cells under circumstances other than standard conditions may be calculated using the exponential Nernst equation (Whitfield and Turner 1981; Millero 2013). Pertinent conditions for oxic seawater are pH ca. 8 ($[H^+] = 10^{-8}$ mol.kg^{-1}), $pO_2 = 0.21$ atm or 358 μmol.kg^{-1}.

4.3.2 Electrochemistry of Metals in Seawater

The prevailing redox potential of oxic seawater, the so-called Eh, can be determined by reading off a voltmeter, the potential generated by an electrode pair consisting of a bare platinum *indicator electrode* (an electrode that acts as a substrate but does not participate in the reaction) and a secondary *reference electrode* of known potential. Seawater, an electrolyte, closes the cell circuit by allowing ion flow to cathode and anode.

Reference Electrodes for Field Use Given the necessity of using cumbersome hydrogen tanks and highly corrosive hydrochloric acid, the standard hydrogen electrode is only used for high precision work in the laboratory and seldom in the field. The sliver/silver chloride and mercury calomel half cells are used instead. These electrodes are self-contained half cells exploiting the low solubility of the chlorides of transition metal. Equilibrium is maintained between the sparse oxidized metal ion and the excess precipitated chloride salt keeping the metal ion concentration stable and, hence, the electrode potential, regardless of the direction of the reaction.

The resulting potential is invariably between 0.5 and 0.6 V and is controlled by the oxygen reduction reactions (Whitfield and Turner 1981; Millero 2013):

$$O_2 + 4H^+ + 4e^- = 2H_2O \qquad Eh = 0.73\ V \tag{4.2}$$

and

$$O_2 + 2H^+ + 2e^- = H_2O_2 \qquad Eh = 0.4\,V \tag{4.3}$$

This potential is remarkably stable despite wide fluctuations in oxygen concentration as is readily confirmed by calculation of the resulting potential through the Nernst equation using environmentally relevant values. Pertinent conditions for oxic seawater are pH ca. 8 ($[H^+] = 10^{-8}$ mol.kg^{-1}), $pO_2 = 0.21$ atm or 358 μmol.kg^{-1}. The tipping point where Eh begins to descend abruptly occurs when oxygen is consumed by microbial activity resulting eventually in a stable reducing environment at Eh values below −0.2 V in which high sulfide concentrations prevail.

The anodic reaction occurring in metal corrosion is generically represented as:

$$nM^0 = M_n^+ + e^- \tag{4.4}$$

where M is a metal that undergoes corrosion. In seawater, these anodic reactions interestingly do not result to any great degree in free metal ions in solution since the higher oxides of most structural metals are very sparingly soluble in seawater and the metal ion is consequently rapidly sequestered as a solid. Hence, since the supply of free metal ions is very low, the reduction of metal ions in solution in the cathodic reaction, theoretically expressed as:

$$M^{+n} + ne^- = M^0 \tag{4.5}$$

does not occur to any large extent but rather the electrochemical reduction of oxygen occurs (Eqs. 4.2 and 4.3). This is because reducible metal ions in seawater solution are present only at vanishingly small concentration, in the nano- (10^{-9}) and picomol.kg^{-1} (10^{-12}) ranges. Alkaline metal ions on the other hand are very abundant with a sodium ion concentration, for example, of 0.469 mol.kg^{-1}, but these metals are not susceptible to reduction at the prevailing Eh. Metallic lithium, sodium or potassium, for instance, when plunged into water will explode in flame. The ion in solution will only return to the metallic state with commensurate energy input. Some lithium cell phone batteries for example have exhibited the unsettling property of exploding in flame.

4.3.3 Metal Corrosion in Seawater

Electrochemical Corrosion

The diversity of metals in use in ocean observing instruments and platforms requires careful consideration of measures to prevent or ameliorate corrosion. Single metal pieces exposed to seawater are subject to galvanic corrosion of their own since different parts of the metal structure in contact with seawater are subject to different ionic environments, such as stagnant water versus moving water, and different degrees of coating with biofilm (see below) thus effectively creating galvanic cells. Contact of dissimilar metals submerged in the seawater medium establishes more powerful *galvanic couples* and corrosion of the least *noble* of the metals in contact ensues rapidly. Steel through-bolts, for example, are commonly used to secure aluminum frames, but such composite structures can easily fail due to corrosion of the aluminum (despite aluminum's self-protecting properties).

Various types of metal corrosion are recognized including generalized surface pitting, stress corrosion cracking, and crevice corrosion. The latter is often due to chemical changes of stagnant waters among and under fouling organisms covering the metal structure (Brown 1969). Dezincification is a particularly insidious form of corrosion attacking brass, alloys of copper and zinc. Since the zinc is selectively lost and the copper remains in place, the piece appears intact but in fact has lost much of its integrity and is fragile and subject to failure.

Sound Brass

A *sound* brass body, in the nautical sense of being undamaged, is elastic and will ring like a bell when struck with a metal piece, or deform if struck too vigorously. An *unsound* piece subject to dezincification and microscopically resembling fine pumice will return a dull sound more like striking wood, or it may indeed shatter if struck too vigorously.

Archeologists take great pains to remove chloride ion (Cl^-) from iron specimens attributing to this ion a role in oxidation of the *ferrous* ion (Fe^{2+}) to *ferric* state (Fe^{3+}). The resulting mineral compound, akagenéite (β-FeOOH) *expands inside the corrosion layers and causes cracking and spalling of the outer surface* (Rimmer and Wang 2010).

Metal abrasion can promote corrosion by exposing fresh metal surfaces to corrosive seawater. Data buoy mooring chains are composed of individual *hot-dip zinc galvanized* links, a coating that greatly improves corrosion resistance of the steel chain. Nevertheless, link to link friction due to wave-induced buoy heave can abrade the protective zinc coating promoting corrosion of the underlying steel. Continued abrasion exposes fresh surfaces to the corrosive seawater. This corrosion is of such concern that chain moorings are replaced at yearly intervals.

Many alloys or single metal formulations of great utility in engineering of marine structures exhibit high corrosion resistance despite low theoretical redox potentials such as aluminum ($E^o = -1.662$) and titanium ($E^o = -0.9$) and various stainless steel and nickel alloys (the light weight and resistance to corrosion of aluminum and titanium make them especially useful in instrument and vehicle manufacture). While these metals and alloys are, in principle, easily corrodible in oxic environments, the oxides formed during initial corrosion impart a thin protective coating to the metal surface impervious to oxygen, thus preventing further corrosion. Aluminum and titanium are self-protective, while the cobalt contained in stainless steel and nickel alloys provides the protective oxide. Nevertheless, these protective coatings may be easily dissolved away if the surrounding electrolyte (seawater) is devoid of oxygen and becomes more acidic as is common, for example, in sediment porewaters or within dense biofouling mats.

Bacterially Mediated Electrochemical Corrosion

Micro-biofouling resulting in the formation of a *biofilm* that coats metallic structures immersed in seawater actively promoting metal corrosion (Videla and Herrera 2005). This biofilm conforms a barrier to oxygen diffusion promoting localized anoxia upon the surface. Under such conditions, heterotrophic marine bacteria transition to anaerobic metabolism using a number of alternate electron acceptors including nitrate, metallic iron and manganese, and nitrate and sulfate as oxidizers of organic matter (Froelich et al. 1979). This increased demand for oxidants leads to the loss of the protective metal oxide coatings exposing the bare metal as the metal oxides in the coating are coopted for bacterial metabolism. Moreover, bacterial metabolism lowers the prevailing pH due to the production of carbonic, sulfuric, and nitric acids which further promote the solubilization of metal oxides and attacks the newly exposed bare metal. Selective or accidental abrasion of parts of the surface can establish redox pairs between the coated and uncoated regions thus promoting anodic corrosion of the sites coated with biofilm relative to the cathodic oxygen reduction occurring at the recently exposed material.

After chloride, sulfate is the most abundant ion in solution in oxic waters. Nitrate, the preferred electron acceptor for facultative anaerobes in the absence of oxygen due to its high energy yield (nitroglycerine and trinitrotoluene explosives attest to this remarkable yield), is present in much lower concentrations and is rapidly exhausted as are the manganous and ferrous ions leaving sulfate, as a low-yield but abundant electron acceptor. Accepted knowledge has been that sulfate-reducing bacteria participate in iron corrosion indirectly through the generation of reducing sulfur compounds such as thiol, mercaptan, and hydrogen sulfide which directly attack the iron surface leading to the coupled cathodic:

$$2H_2S + 2e^- = 2SH^- + H_2 \tag{4.6}$$

and anodic reactions:

$$Fe^0 \rightarrow Fe^{2+} + e^- \tag{4.7}$$

The more soluble *ferrous* ion (Fe^{2+}) then becomes available to other facultative anaerobes that complete its oxidation to the insoluble *ferric* form (Fe^{3+}) which precipitates as carbonates, oxides, and oxihydroxides.

Under anoxic conditions, such as pipeline interiors and sediment-embedded structures, microbiologically influenced corrosion (MIC) proceeds at economically significant rates incompatible with the slower indirect reactions proposed above. Enning et al. (2012) and Enning and Garrelfs (2014) have now demonstrated that sulfate-reducing bacteria can directly oxidize metallic iron in carbon steel through a lithotrophic process:

$$4Fe^0 \rightarrow 4Fe^{2+} + 8e^- \tag{4.8}$$

mediated through sulfate reduction:

$$8e^- + SO_4^{2-} + 9H^+ \rightarrow HS^- + 4H_2O \tag{4.9}$$

The authors propose that electron flow for the sulfate reduction reaction occurs across the conductive biogenic mineral crust (FeS, $FeCO_3$, Mg/$CaCO_3$) that is deposited on the metal substrate in the process.

Following primary oxidation to the ferrous state, characteristically black, numerous bacterial groups are capable of carrying out the second oxidation step

$$Fe^{2+} \rightarrow Fe^{3+} + e^- \tag{4.10}$$

resulting in the deposition of highly insoluble reddish ferric oxides and very sparingly soluble oxihydroxides.

RMS Titanic Rusticles

Spectacular video imagery of the wreck of the Royal Mail Ship Titanic obtained by Ballard and collaborators in 1986 aboard the research submersible Alvin deep in the north Atlantic highlighted the discovery of *rusticles*, thick coatings and chords of a mineral/biofilm accretion rich in reddish ferric oxihydroxides hanging from rails and other surfaces. Projections of total collapse of the wreck within this current century highlighted in the public eye the role of bacteria in mediating metal corrosion. In fact, a new bacterium species named *Halomonas titanicae* was discovered and described by researchers from Spain and Canada in rusticles samples subsequently recovered from the wreck. The metabolic strategy for iron oxidation of *H. titanicae* remains to be elucidated. Video imagery of the wreck and its rusticles can be viewed at https://oceantoday.noaa.gov/titanicwrecksite/. Accessed September 16, 2017.

4.3.4 Corrosion Protection

Preventive Corrosion Protection

In consequence of the multiple pathways for marine corrosion and its severe economic impact, a primary precaution for those operating electromechanical equipment in the seawater is its prevention. Avoidance of galvanic pairs with direct contact of dissimilar metals immersed in the seawater electrolyte is foremost among these precautions. Protective measures can be as simple as the use of nylon washers between metal parts. Shoulder washers incorporate the traditional circular flat washer continuing to an orthogonal cylindrical inset lining thus isolating through-bolts from contact with the piece.

Additional to insulation of potential galvanic pairs, certain rules of thumb may be followed which help to ameliorate corrosion problems (Brown 1969; Horne 1969). Metal selection for small but important structural components such as bolts, rivets, or hinges should be such that they constitute cathodes with reference to the overall structure in order to prevent their corrosion which can otherwise result in catastrophic failure of the assembly. The structural material constituting the anode will tend to corrode but corrosion will be generalized and structural integrity will be extended. Small cathodic pieces may be painted or embedded in epoxy to seal out the seawater electrolyte interrupting the electrolytic redox cell preventing corrosion of adjoining anodic couples. Paint or polymeric protective coatings may be used but may prove to be counterproductive if used unwisely. For many applications, large structural parts composed of self-protecting metals may be left uncoated since otherwise the coating might fail at points of greatest stress or flexion, points that may be critical to structural integrity and where corrosion will concentrate. Slow generalized surface corrosion may prove preferable to intense localized *hot-spot* corrosion.

Cathodic Protection

Cathodic protection is a technology designed to render metal-containing structures impervious to electrolytic corrosion by providing an alternate electron source for oxygen reduction (Brown 1969; Horne 1969). Since corrosivity increases as the redox potential of the individual metal/seawater half cells diverges, it is also standard practice to affix an easily corroded metal, usually zinc, to the structure to be protected. The zinc here constitutes a *sacrificial anode* designed to be lost in place of the protected *cathodic* structure through which the electrons from the corroded anode flow to satisfy the electron demand of dissolved oxygen as per reactions 4.2 and 4.3.

Larger structures such as ocean towers may be provided with active cathodic protection using DC powered cathodes. For smaller scale applications such as instruments and vehicles, *sacrificial anodes* provide an economical solution. Alloy formulations for use as sacrificial anodes are designed to purposely corrode when affixed to the structure of interest by creating a galvanic pair thus protecting instruments and platforms containing metal parts. In practice, various solid metal alloy shapes are fitted to the application. Solid cylinders or flat ingots may be bolted to carousels or buoy infrastructure; toroidal shapes may be clamped to instrument cages; *finger* anodes inserted in the cooling manifolds are common to raw seawater cooled diesel engines. Smaller units distributed throughout the structure prove to be more effective than massive single units since current flow is dispersed avoiding corrosion hotspots (Fig. 4.1).

Fig. 4.1 Sacrificial zinc anode

Alloy formulations are available for a range of applications from freshwater to seawater progressing from more friable magnesium alloys through mid-range aluminum, to zinc, specified for practical seawater salinity above 30. Sacrificial anode endurance depends largely on current speed. Care must be taken to protect the junction between anode and subject material with a nonconductive coating such as an epoxy sealant.

Cathodic protection relying on sacrificial anodes is known as passive protection. Active or *impressed current* cathodic protection may prove advantageous for larger structures such as towers. A direct current electrical (DC) potential is applied to inert metal cathodes connected to the surface to be protected. Anodes may be graphite or high silicon iron or may consist of platinized copper or copper core titanium usually shaped as 12 mm diameter rods (Bahador 2014). Reduction of molecular oxygen to a hydroxide, hydrogen peroxide, and/or, water at lower potentials completes the redox cell. With higher DC potentials impressed to the anode (in addition to oxygen reduction) may result in gaseous hydrogen generation during anodic reactions of the form:

$$O_2 + 2H_2O + 4e^- = H_2 + 2OH^- \tag{4.11}$$

Or, at sufficiently high potentials, may entail reduction of the abundant chloride ion to the molecular gas:

$$2Cl^- = Cl_2 + 2e^- \tag{4.12}$$

The resulting chlorine gas is a potent biocide and can be generated purposely for active biofouling control (see below). Since DC potentials up to 100 V may be applied, a structure may be protected using a single anode connected through a wire harness distributed throughout the structure. For large structures such as ship hulls, computer controlled networks of multiple active anodes paired to Ag/AgCl reference electrodes allow optimized DC power distribution to address localized corrosion-inducing potentials.

4.4 Biofouling of Ocean Observing Instruments and Platforms

Macro-biofouling is the accumulation of macroscopic sessile marine organisms on submerged surfaces. Ocean observing instruments and platforms immersed in seawater are immediately subject to biofouling which can affect both instrument and platform performances. Instrument sensors may be mechanically and optically occluded leading to gradual degradation of the signal and eventual sensor failure. Platforms may similarly be fouled leading also to decreased performance and failure. Moving parts may be jammed and increased drag is experienced by mobile platforms. Small marker buoys may even sink under the weight of the aggregated fouling community. Anticorrosion coatings can also be destroyed by fouling organisms.

Biofilm or micro-biofouling, a surface coating consisting of various types of microorganisms embedded in an aqueous mucous sheath, is the precursor to macro-biofouling. When a clean substrate is immersed in seawater, a film of dissolved organic material (ubiquitous in the marine environment) adsorbs upon the surface facilitating settlement of pioneering bacterial colonizers (Videla and Herrera 2005). These in turn promote aggregation of other microorganisms and the formation of the mucous film via *quorum sensing*, a chemical communication system consisting of the secretion and detection of small signaling molecules known as *autoinducers* which in turn promote secretion of extracellular polysaccharides (40–95%), proteins, DNA, lipids, and other substances. Microorganisms embedded within the biofilm exhibit enhanced resistance to antibiotics and biocides.

Bacterial biomass by far dominates biofilm colony composition with prominent presence and role for the Proteobacteria. *Roseobacter* and *Alteromonas*, members of the Proteobacteria associated to *marine snow*, are among the primary early colonizers. Pennate diatoms, notably the genera *Nitzschia, Amphora, Cylindrotheca*, and *Navicula*, constitute the second most abundant group of microorganisms. Cyanobacteria of the genera *Synechococcus* and *Oscillatoria*, green flagellates, heterotrophic microflagellates, and dinoflagellates may also constitute part of the biofilm community.

Extracellular polysaccharides have been isolated from culture of a number of marine bacterial genera including *Vibrio, Pseudomonas, Lactobacillus, and Weissella*. Most produce heteropolymers containing glucose, galactose, rhamnose, mannose and other pentoses, hexoses, hexsosamines, and uronic acids, the so-called mucopolysaccharides or glycosaminoglycans. These compounds constitute the gelatinous matrix of the biofilm. Autoinducers are principally of the acylhomoserine (AHL) family of compounds. Secretion of AHL into the medium promotes phenotypic changes of otherwise free-living bacteria which enhance bacterial adhesion, polysaccharide biosynthesis, colony formation, and enhanced surface motility (Di Donato et al. 2016; Steinberg et al. 2002). Free-living flagellate forms are known to lose the flagellae upon settling.

Mature fouling assemblages host animals of most phyla including the simpler sponges and bryozoans; coelenterates such as hydroids and corals; segmented serpulid tube worms; crustaceans such as barnacles, pedunculated goose barnacles (cirripeds); various species of decapods including shrimps, crabs, and lobsters; mollusks such as mussels and oysters; echinoderms; and lower chordates (tunicates), together with a diversity of macroalgae. All are characterized by a meroplanktonic lifestyle, in which larvae are planktonic but metamorphosis precipitates settling onto a substrate followed by development of the adult life form. Community composition and diversity depend in large part on geographic location. Arctic assemblages for example differ substantially from their tropical counterparts, and fouling communities in nearshore waters similarly differ from those developing upon surfaces moored, floating or traversing the surface of the deep ocean. Pelagic goose barnacles *Lepas sp.*, for example, are emblematic within the reduced diversity of offshore fouling assemblages. Assemblages also differ as the fouling process advances with simpler smaller organisms serving as initial pioneer colonizers followed by larger more complex organisms. Bryozoans are among the early colonizers followed by serpulid tube worms, mollusks, and cirripeds. Sessile organisms that adhere firmly to the surface precede motile free-living benthic organisms such as shrimps, crabs, or starfish since a suitably complex environment must be present to host the latter.

Surface colonization by multicellular eukaryotes is modulated by chemical cues including both inducers, which promote colonization, and deterrents which discourage it (Steinberg et al. 2002). Inducing cues may be released into the water or may be surface bound. In addition to ACH, small peptides, larger proteins, carbohydrates and polysaccharides, glycerolipids, and fatty acids have all been found to induce settlement of a variety of meroplanktonic taxa. Noxious natural substances, such as halogenated furanones from macroalgae and sponge metabolites or turpentine used to preserve wooden pilings, are deterrents to the settling of meroplanktonic larvae. Natural deterrents are usually nonpolar water insoluble compounds concentrated in the external tegument of benthic organisms.

Substrate colonization occurs through biological succession usually from the simpler organisms to the more complex. Thus, micro-biofouling progresses from bacteria to unicellular autotrophs to unicellular heterotrophs. Since dissolved organic matter rapidly (within hours) adsorbs to clean surfaces immersed in seawater, a suitable substrate for bacterial copiotrophs appears, which organisms, upon colonization prepare the way for less exacting oligotrophic bacteria. These two groups in turn create the conditions for colonization by autotrophs (diatoms) by creating an environment rich in inorganic nutrients, the sum of which can now sustain heterotrophic microorganisms such as ciliates and flagellates. Mature biofilms set the stage for colonization by fast-growing macroorganisms, mainly hydroids, bryozoans, serpulids, and ascidians, followed later by slower growing mollusks (Railkin 2004; Delauney et al. 2010). Motile organisms are the last to appear once a complex substrate is available. Such climax states may be attained within months to years depending on the climate and nutrient supply. Juvenile crabs, spiny lobster, and gobiid fish are regularly seen during annual refurbishing of the CariCOOS MapCO2 buoy deployed in a tropical coral reef environment.

Of concern to ocean observing is the deterioration of instrument response due to biofouling, the rate of which depends greatly on the deployment site and deployment mode. Performance of identical instruments deployed in the high productivity waters at the confluence of the Patuxent River and Chesapeake Bay (Solomons) and in the low productivity environment of Kaneohe Bay in Hawaii during ACT field tests differed widely as discussed previously with rapid fouling occurring at the former site. Vehicles in motion and their instrument payload are subject to differential fouling in accordance to localized turbulence and water velocity. Surfaces subject to uninterrupted laminar flow discourage fouling while recondite, high turbulence areas favor increased fouling. Biofouling is minimal in the deep sea, one of the reasons for *parking* the profiling Argo floats at 2 km depth between ascents.

Stringent measures are consequently taken to minimize biological fouling of vehicles and sensors deployed in near-surface waters, particularly for extended deployment.

Antifouling (AF) technologies are directed toward the mechanical and biocidal prevention of adhesion of marine organisms to working surfaces. Application of either biocide-containing coatings or silicone or fluorinated polymer *fouling release* coatings (Salta et al. 2013; Delauney et al. 2010) follows standard practices developed in the maritime and offshore petroleum industries. Biocide coatings may be based on toxic metal oxides or complexes such as the various copper or zinc formulations available commercially. Tributyltin was once in widespread use but its severe environmental effects have resulted in tight restrictions on its use. The organometallic complex zinc pyrithione and the photosynthesis-inhibiting thiazine irgarol are sometime used in conjunction with copper formulations. Self-polishing or *ablative* paints are designed so that the coating slowly wears off while the structure is in movement or subject to moving water. Hard paints, usually epoxy based, contain high metal concentrations and do not wear off. However, the metal eventually becomes oxidized and loses efficacy. In either case, recoating becomes necessary at temporal ranges of months to years depending on the efficacy of the coating and the challenge of the environment.

While many such formulations are effective in preventing macro-biofouling, less success has been achieved in preventing micro-biofouling. Hydrophobic surfaces such as the silicone release coatings have been found to be ineffective in preventing bacterial adhesion. Diatoms, particularly the genus *Amphora*, have proven to be quite resistant to both the biocidal and fouling release coatings but susceptible to copper ablative formulations when in movement during dynamic tests.

For long-term deployment, metallic copper, an effective biocide, is used to sheath instrument housings in the form of commercially available copper foil with an adhesive backing for ease of application. Other more sensitive and intricately fabricated exposed parts may be surrounded by copper mesh (Fig. 2.14). Antifoulant cartridges containing tributyltin (TBT), an organometallic biocide, are used in flow-through conductivity cells to combat fouling. Optical instruments may be fitted with mechanically operated rubber wipers to periodically remove fouling from optical windows or copper shutters may be installed opening only for measurements. Active means of sensor protection include chlorine gas sterilization through chloride ion reduction at electrolytic anodes in close proximity to the subject sensor, or by periodically delivering small aliquots of acid to the sensor region (Delauney et al. 2010). Other energy intensive methods such as periodic heating and increases of water current flow to prevent larval settlement have been proposed and tested (Galler 1969) but are not in use for ocean observing at present.

Ocean gliders pose a unique case of vulnerability to biofouling due to their extended missions and repeated near-surface operation. Rubber-like nonstick materials have performed well in extend glider operations. The historic Atlantic Ocean crossing in 2009, a 4600 mile journey, performed by a Rutgers University RU Slocum glider (Lobe et al. 2010) provided a test bed for such surfaces in ocean observing operations. Coated surfaces suffered minimal fouling, while uncoated surfaces were subject to *Lepas* fouling. As expected fouling was most severe in semi-protected low-turbulence parts of the vehicle which led, at one point, to loss of directional control due to fouling of the rudder hinge.

4.5 Mechanical Abrasion of Ocean Observing Platform Components

High energy environments can cause abrasion of anchoring and electrical systems. Chain link abrasion due to wave and current action, added to rapid oxidation of the recently exposed metal, can cause catastrophic failure of buoy anchor systems resulting in loss of the anchoring system thus setting the buoy adrift, a significant safety concern. Best practice is consequently the periodic replacement of the entire anchor chain used for coastal buoys.

Strong wind, current, and wave action can cause *strumming* (rapid vibration) of electrical wiring that can lead to component failure. As practiced in naval, aeronautical, and automotive engineering, electrical wires are bunched together in wiring harnesses wrapped in protective tape or plastic coverings. Additional *chafing gear* such as flexible tubing or helical sheaths may be added at critical contact sites between metal and the wiring harness, especially if in contact with metal edges. Wiring harnesses such as those leading from submerged or tower-mounted instruments to the buoy instrument bay are firmly secured using nylon cable ties throughout the cable run.

Shore-crossing power and communications cable are in these days generally tunneled underground using directional boring, in large part due to environmental restrictions, but in good measure due to the possibility of cable failure resulting from abrasion caused by high energy breaking waves, especially along rocky shores (Pierce and Romanelli 1969).

References

Bahador A. Cathodic corrosion protection systems: a guide for oil and gas industries. NY: Elsevier; 2014. ISBN: 9780128003794. 492 p.

Brown BF. Corrosion. In: Myers JJ, Holm CH, McAllister RF, editors. Handbook of ocean and underwater engineering. Section 7: Materials and testing marine corrosion, boring and biofouling. New York: McGraw Hill; 1969. pp. 3-4 to 3-30.

Delauney L, Compere C, Lehaitre M. Biofouling protection for marine environmental sensors. Ocean Sci. 2010;6:503–11. https://doi.org/10.5194/os-6-503-2010. www.ocean-sci.net/6/503/2010/. Accessed 15 Sept 2017.

Dewan A, Ay SU, Nazmul Karim M, Beyenal H. Alternative power sources for remote sensors: a review. J Power Sources. 2014;245(2014):129–43.

Di Donato P, Poli A, Taurisano V, Abbamondi GR, Nicolaus B, Tommonaro G. Recent advances in the study of marine microbial biofilm: from the involvement of quorum sensing in its production up to biotechnological application of the polysaccharide fractions. J Mar Sci Eng. 2016;34:14. https://doi.org/10.3390/jmse4020034.

Enning D, Garrelfs J. Corrosion of Iron by sulfate-reducing Bacteria: new views of an old problem. Appl Environ Microbiol. 2014;80(4):1226–36.

Enning D, Venzlaff H, Garrelfs J, Dinh HT, Meyer V, Mayrhofer K, Hassel AW, Stratmann M, Widdel F. Marine sulfate-reducing bacteria cause serious corrosion of iron under electroconductive biogenic mineral crust. Environ Microbiol. 2012;14(7):1772–87. https://doi.org/10.1111/j.1462-2920.2012.02778.x. PMCID: PMC3429863.

Froelich PN, Klinkhammer GP, Bender ML, Luedtke NA, Heath GR, Cullen D, Dauphin P, Hammond D, Hartman B, Maynard V. Early oxidation of organic matter in pelagic sediments of the eastern equatorial Atlantic: suboxic diagenesis. Geochimica et Cosmochimica Acta. 1979;43(7):1075–90.

Galler SR. Boring and fouling. In: Myers JJ, Holm CH, McAllister RF, editors. Handbook of ocean and underwater engineering. Section 7: Materials and testing marine corrosion, boring and biofouling. New York: McGraw Hill; 1969. pp. 7-12 to 7-19.

Horne RA. Marine chemistry. New York: Wiley-Interscience; 1969. 568 p.

Lobe H, Haldemann C, Glenn SC. *ClearSignal* coating controls biofouling on the Rutgers glider crossing. Sea Technology. 2010. https://wwwsea-technologycom/features/2010/0510/clearSignal_coating_controlshtml#top. Accessed 14 Sept 2017.

Millero FJ. Chemical oceanography. 4th ed. Boca Raton: CRC Press Taylor & Francis Group; 2013. 571 p. ISBN 9788-1-4665-1249-8.

Pierce GA, Romanelli RP. Cable installation and Repair. In: JJ Myers, CH Holm, RF McAllister (Eds.). Handbook of ocean and underwater engineering. Copyright by North American Rockwell Corporation. New York: McGraw Hill; 1969. pp. 5–29 to 5–51.

Railkin AI. Marine biofouling: colonization processes and defenses. Boca Raton: CRC Press; 2004. 300 pp. ISBN-13: 978-0849314193 ISBN-10: 0849314194.

Rimmer M, Wang Q. Assessing the effects of alkaline desalination treatments for archaeological iron using scanning electron microscopy. Br Museum Tech Bull. 2010;4:79–86.

Rudnick DL, Davis RE, Eriksen CC, Fratantoni DM, Perry MJ. Underwater gliders for ocean research. Mar Technol Soc J. 2004;38(2):73–84.

Salta M, Wharton JA, Blache Y, Stokes KR, Briand J-F. Marine biofilms on artificial surfaces: structure and dynamics. Environ Microbiol. 2013;15(11):2879–93.

Statsewich H, Weingartner T, Grunau B, Egan G, Timm J. A high-latitude modular autonomous power, control and communication system for application to high-frequency surface current mapping radars. Mar Tech Soc J. 2011;45(3):59–68.

Steinberg PD, de Nys R, Kjelleberg S. Chemical cues for surface colonization. J Chem Ecol. 2002;28:1935–51.

Vanýsek P. Electrochemical series. In: Haynes WM, editor. CRC handbook of chemistry and physics. 91th ed. Boca Raton: CRC Press/Taylor & Francis Group; 2010. ISBN-13: 9781439820773. pp. 8-20–8-29.

Videla HA, Herrera LK. Microbiologically influenced corrosion: looking to the future. Int Microbiol. 2005;8(3):169–80.

Whitfield M, Turner DR. Sea water as an electrochemical medium. In: Whitfield M, Jagner D, editors. Marine electrochemistry. Chichester: Wiley; 1981. p. 1–63.

Chapter 5
Signal Conditioning, Data Telemetry, Command Signaling and Platform Positioning in Ocean Observing

Abstract *Data telemetry* is the transmission of data from remote observing platforms. Transmission from control station to platform, an additional ability afforded by two-way communication (Tx/Rx) systems, allows *command signaling* for adjustment of mission parameters. Autonomous navigation requires the capability to receive navigational positioning data from satellite systems such as GPS or from underwater acoustic beacons, or the use of onboard inertial navigation systems. Data from distributed sensor networks is transmitted through space and through the atmosphere via radio waves, through the water via acoustic means, through cables as current pulses, or as optical pulses through fiber-optic waveguides. Since seawater is largely opaque to radio signals, submerged ocean observing applications relying on satellite and cellular network communications for transmission must have alternate means of communication between the submerged platform and a surface relay platform, or must periodically reach the surface for data transmission. Data are transmitted via modulation of a carrier wave. Signal conditioning consists of amplification of the initial signal, signal filtering to minimize electronic noise, digitization, and eventual carrier signal modulation in preparation for transmission.

Keywords Signal conditioning · Analog to digital conversion · Carrier modulation · Data telemetry · Command signaling · Platform positioning

5.1 Data Signal Conditioning for Ocean Observing

Most sensor signals, such as variations in electrical current, resistance, voltage, or frequency arising as the primary response of an electronic transducer to physical forcing, are within the analog realm where variations are normally smooth and continuous.

The original version of this chapter was revised. A correction to this chapter can be found at https://doi.org/10.1007/978-3-319-78352-9_9

J. E. Corredor, *Coastal Ocean Observing*,
https://doi.org/10.1007/978-3-319-78352-9_5

Telling Analog from Digital

Analog and digital can easily be differentiated by considering the data processing of an old-style reversing thermometer. Then, the researcher would read out the analog signal of the mercury meniscus height against the calibrated scale through a magnifying glass and, putting the data (a set number of digits) to paper, would, in the act, *digitize* the reading. Two digits were directly read off the scale and, with a sharp eye, a third digit could be subjectively interpolated.

Similarly, early electronic instrument readout such as CTD and XBT temperature or salinity readings consisted of an analog stylus trace on graph paper. It was necessary to digitize the readings for both temperature and depth by reading out against the graph paper line grids graduated in meters and degrees centigrade or parts per thousand salinity. The last significant digit was similarly estimated.

Preparing electronic sensor signals for storage and transmission takes the name of signal conditioning. Initially, the tenuous sensor signal is electronically amplified by applying the data signal, as current or voltage, to the gate of a power transistor across which a high power carrier signal is modulated thus achieving amplification of the initial signal up to about a hundredfold. This output is then digitized and filtered for storage and transmission. Electronic digitization called *quantization* is carried out by electronic modules known as analog to digital converters (ADC). Central to the ADC is the sample-and-hold circuit whereby the signal voltage is removed from the sensor circuit over the interval of the preestablished instrument time set (the clock speed), stored in a capacitor and, from there, sampled by *comparator* circuits to determine the appropriate voltage bin. Some comparators sample to an array of voltage bins, others ramp through a preestablished number of voltage bins. In any case, the number of possible bin determines the resolution of the ADC.

Quantization and Data Precision

Quantization necessarily causes error due to the rounding off procedure which truncates an essentially infinite number. Quantization error however decreases as the resolution of the circuitry increases. An initial example expressed in decimal notation is instructive: For an error of one in one hundred (1:100), a resolution of two digits suffices (00–99) and a range of one hundred discrete values will represent the analog span observed. If digitization resolutions allow reducing the quantization error to one in one thousand (1:1000), three digits are required (000–999) to represent the data set. It is to be noted that the increased resolution of one digit makes the size of the data set increase by a factor of 10. In binary, rather than decimal notation, these resolutions are measured in bits. For perspective, a 10 bit number covers a decimal range from 0 to 1024 (2^{10}). Since AD converters today can perform quantization ranging up to 24-bit resolution, more than 16 million (2^{24}) possible discrete decimal digital bins may be achieved!

Resolution and the size of the data set likewise increase as sampling frequency of the ADC increases. Converter sampling rate is commonly quoted as samples-per-second (sps). In order to capture the variability of the signal, the sampling frequency must be at least half the period of the highest frequency to be sampled. Current technology allows sampling rates up to 300 Msps allowing reliable sampling of frequencies up to about 150 MHz, precision rarely required for environmental applications. In summary, as resolution in time and in sensor response readout increases, precision increases but the data set becomes larger. This tradeoff has prompted a continuing race between data generation rate and data storage and handling capacity. Advances in circuit miniaturization and technology development have fortunately until now relieved this bottleneck.

An effective environmental sensor is one capable of measuring a signal across its natural analog span at numerical resolution commensurate with the application at hand. The time-referenced analog output of an effective sensor will result to be graphically representable as a smooth succession of peaks and valleys of different frequencies and amplitudes. In practice, single digit resolution (0–9) is considered poor, so sensor output values are usually quoted with a precision of two or three decimal digits and only rarely (and with sufficient justification) at greater resolution.

Fourier theory proposes that any such time series can be represented as a composite of superimposed sinusoidal variations. Breaking down a signal into component frequencies throughout this spectrum allows a simple solution to ameliorate signal noise; the application of frequency cutoff filters. Rapid positive and negative fluctuations superimposed upon a long smooth wave representing a low frequency signal can be removed by cutting off variations at frequencies above that low frequency band of interest, a procedure known as the application of a *low band pass filter*. A high frequency signal superimposed upon low frequency noise may be conditioned by filtering out only low frequency variations constituting a *high band pass filter*. *Cutoff band pass filters* can isolate the band sought or alternately blank out a noisy band in an otherwise informative spectrum. Early electronic filters used circuits composed of capacitors, induction coils, and resistors to dampen out the unwanted signal. Various piezoelectric microelectromechanical devices are in use today.

Electronic Noise

Noisy data may require electronic filtering. Use of the term *noise* arises of course from the field of audio reproduction and the analog audiophile recognizes electronic noise as the hiss of the needle or the tape, the scratches on the vinyl, or the gaps on the tape. When the noise surpasses the signal, the recording can no longer be distinguished. Since the *signal-to-noise ratio* dictates the resolution, very noisy signals must be filtered prior to quantization.

The Global Telecommunications System (GTS) is a major component of transmitting global meteorological data, consisting of both in situ and satellite observations. This data is collected by a number of organizations which archive and further process the data. In the US, the National Centers for Environmental Prediction

(NCEP) collect this GTS data and format it into BUFR (Binary Universal Form for the Representation of meteorological data) for their processing needs. After processing, the BUFR is transmitted to the National Center for Atmospheric Research (NCAR) for archival and additional processing into the International Maritime Meteorological Archive (IMMA) format. The IMMA format has been adopted by many organizations as the preferred format for marine observations. The data consists of basic observations taken from ships, buoys, C-MAN (Coastal Marine Automated Network), and tide-gauge stations. Observations may include air and sea surface temperature, wind direction/speed, waves, sea level pressure, etc.

All such transmissions are performed by superimposing digital signal data upon established radio or microwave carrier frequencies through electronic modulation, an integral part of the signal conditioning process.

Radio AM and FM

Readers are probably acquainted with analog audio radio amplitude modulation (AM) and frequency modulation (FM). In the former, the amplitude of a high power carrier wave passing across a transistor at a fixed frequency is modulated by applying the small microphone signal to the gate acting as an electronic valve to modulate the amplitude of the high power carrier wave. The carrier signal is applied to the transistor source and transmitted from the drain to the antenna. In the more complicated FM circuit, the amplitude is invariant but the frequency is modulated. Since it is the frequency that is modulated, a spectral band of some width is required. For commercial analog FM radio transmission in the USA between 87.8 MHz and 108.0 MHz, for example, individual station band width is set at 0.2 MHz allowing only 101 stations across the effective 20.2 MHz FM band.

Digital data transmission makes use of frequency modulation schemes. Some internal data recording CTD instruments, for example, use *frequency shift key* modulation (FSK) to generate hexadecimal data (a numeral system with a 16 digit base) for rapid instrument data download (the *ASCII* representation of hexadecimal numerals is a set of numbers (0–9) and letters (A-F) adding up to a total of 16 characters). More commonly, data is transmitted as binary code in sequences of ones and zeros. For such applications, a four-key FSK scheme (i.e., four different adjacent keyed frequencies) can be used to represent two bits per key (00, 01, 10, 11) thus increasing data density by a factor of 2.

Carrier wave phase is an additional property of electromagnetic radiation that can be exploited for data transmission in addition to amplitude and frequency. Various schemes are available for *phase shift key* (PSK) modulation. Amplitude shift key (ASK) modulation is possible, but AM radio bands are notoriously noisy and are avoided for radio transmission in data telemetry. Nevertheless, some optical data transmission modes through fiber-optic cable do implement ASK. As the object of data transmission is to transmit data with a minimum of error while maximizing data throughput, many combinations of such schemes are used. As the number of keying schemes and the number of frequencies available increase so also does the data throughput rate. It must be remembered that data throughput rate varies enormously and such high throughput rates are only necessary for very demanding applications such as streaming video or audio or synoptic satellite imagery. Many applications such as hourly buoy instrument data rely on simple GSM text messaging (see below).

5.2 Electromagnetic Data Transmission for Coastal Ocean Observing

Electronic satellite data transmission and relay functions using electromagnetic means are performed today at frequencies between 300 MHz and 300 GHz, in the high band radio and microwave spectra. These bands are preferred since they allow greater data density, directionality, and smaller antenna dimensions. Greater innate data density is due to the higher frequency. In addition, while VHF-UHF may be reflected by the ionosphere, the higher frequency microwaves reliably penetrate this atmospheric layer and are thus essential for satellite communications. Microwaves propagation is strictly line-of-sight. Obstacles between the Tx/Rx pair will impede communication. The low wavelength of microwaves (0.1–100 cm) allows the use of moderately sized parabolic reflectors which impart directionality to the resulting radiation beam greatly increasing the useful range. Omnidirectional antennae are used for transmission from the instrument platform source to the receiving antenna. Microwave relay such as from a satellite to an earth station is achieved using highly directional feed horn or parabolic reflector arrays.

Microwave *oscillators*, electronic devices that convert direct current to microwave frequencies, generate the carrier wave at the desired frequency. The first oscillators developed for radar transmission during the Second World War were the klystrons, bulky vacuum tube device. Much more compact and reliable solid-state devices incorporating field effect and junction-bipolar transistors are in use today (Khanna 2006).

5.2.1 *Satellites in Terrestrial Orbit as Communication Platforms*

Robust data telemetry, platform positioning, and vehicle tracking are essential for sustained, autonomous ocean observing. Cellular networks now provide effective low-cost data telemetry but range limited to about 35 km restricts this application to nearshore areas. Satellites in earth orbit on the other hand offer widespread data coverage at moderate cost. The era of satellite communications dawned with the 1962 launching of the TELSTAR satellite that for the first time provided an extended global telephone, FAX, and television link. Today several state-sponsored and commercial satellite communication systems provide data links (including positioning data) from surface *Data Collecting Platforms* (DCP) throughout the world oceans. DCP data can in most cases be received by the operator through internet links.

The ARGOS Data Collection and location System (DCS) is the precursor to all satellite data relay and platform location services. Operational since 1978, ARGOS instruments aboard US, European, and weather satellites in polar orbit provide both a data transmission service and platform location based on the Doppler Effect with precision ranging from about 150 m to 1000 m. It is an older, less precise, but effective alternative and supplement for platform location to the newer, highly precise GPS system (see below). However, ARGOS provides data relay for platforms equipped with GPS receivers allowing GPS precision as needed. Initially emplaced and operated jointly by the US National Atmospheric and Space administration (NASA), the US National Oceanic and Atmospheric Agency (NOAA), and the French Centre National d'Etudes Spatiales (CNES) and joined today by the European Organization of the Exploitation of Meteorological Satellites (EUMETSAT) and the Indian Space Research Organization (ISRO), ARGOS instruments operate aboard NOAA and EUMETSAT polar orbiting weather satellites in sun synchronous low Earth orbits at altitudes below 900 km. ARGOS-3 transmitters (currently in their third generation) aboard surface platforms operate at a frequency of 401.65 MHz in the lower UHF radio band. Radiated power is about 5 W, in bursts of less than 1 s at intervals of 45–200 s allowing the use of low power sources such as batteries and solar panels (IOC/WMO DATA BUOY CO-OPERATION PANEL, 2011).

NOAA operates a *Geostationary Operational Environmental Satellite* (*GOES*) constellation of satellites in equatorial geostationary orbit at an altitude of 35,786 km. Two primary operational satellites cover the Atlantic Ocean to the east and the Pacific to the west of the USA. *GOES East* orbits directly above South America and *GOES West* orbits above the eastern Pacific Ocean. GOES satellites' primary mission is to support weather forecasting, to track severe storms, and to assist in meteorological research. However, essential to their meteorological mission is the provision of a communication link for surface-based environmental observation platforms. The GOES *Data Collection System* (DCS) offers robust data relay for ocean observations from diverse autonomous platforms located in the Eastern and Central Pacific Ocean and throughout most of the Atlantic Ocean. Since the number of DCPs is dictated by the number of available channels, assigning narrow temporal communication windows to individual DCPs allows increasing the number of platform communications. Platforms usually collect data at higher programmed rates but transmit data packets at lower rates, typically hourly for operational ocean observing surface buoys. The DCS aboard GOES receives data from more than 10,000 DCPs. DCP data is received at GOES East through channels at 468.250 MHz (domestic) and 402.200 MHZ (international). DCP data is then relayed to ground stations via the microwave radar L band at 1694.500 GHz. DCP interrogation is also available at 2034.9000 MHz (Nestlebush 1994). Similarly, in support of its meteorological mission, the METEOSAT system operated by EUMETSAT supports DCP transmission in its area of coverage. Two operational geostationary satellites centered above the equatorial Eastern Atlantic and the Indian Oceans provide 11 international channels (402.0355–402.0685 MHz) and 234 regional channels (402.0685–402.435, MHz) in the HF band (EUMETSAT 2013). The Japanese Himawari 8 geostationary weather satellite operated by the Meteorological Satellite Center of the Japan meteorological Agency receives DCP data through its international channel at 402.0–402.1 MHz and domestic channel 402.1–402.4 MHz. Data is transmitted to ground segments in the radar Ka-band at 18.1–18.4 GHz.

The GLOBALSTAR system is a commercial constellation of satellites in low earth orbit at an altitude of about 1400 km with a 114 min orbit period. Following degradation of communications in the S-band amplifiers of the first generation, a second generation of 24 satellites in 8 orbital planes with improved communications capability was placed in orbit and is operational as of 2013. Forty ground receiving stations complete the communications network. For data transmission from remote locations, GLOBALSTAR offers uplink communications to the satellite within 1610 and 1625 MHz in the MF radio band and downlink communications to receiving stations within 2483.5–2500 MHz microwave band. GLOBALSTAR electronic modules (*databoards*), in addition to providing a data link, provide GPS reception allowing platform tracking through this satellite network. Coastal area coverage is provided for the Americas, Europe, Australia and Japan.

The IRIDIUM system is also a commercial satellite constellation that provides global voice and data transmission. The 68 operational satellites in this constellation orbit an altitude of approximately 780 km. In contrast to the GLOBALSTAR constellation which exhibits orbits inclined at 52°, IRIDIUM satellites use polar orbits allowing coverage within the Polar Regions. With 66 satellites in 6 orbital planes, IRIDIUM operates user links for DCPs within 1618.85 and 1626.5 MHz in the HF radio band. A second generation of 66 operational satellites operating in the L radar band is expected to be in place by 2020. IRIDIUM satellites communicate with four neighboring satellites within the constellation in band between 22.55 and 23.55 GHz allowing flexible communication relays.

5.2.2 Cellular Network Data Transmission for Coastal Ocean Observing

Platforms generating low data density such as ODAS buoys communicate over cellular networks using the Global System for Mobile communications (GSM) system, an international standard data transmission technology operating in the bands between 300 and 1900 MHz. Since higher power output (up to 20 W) than that allowed for cell phones is allowed for *GSM radios* operating at these frequencies, transmission range is extended. Maximum GSM range is quoted at about 35 km. Cellular communications networks are today fully digital, in large part, to facilitate data transmission. The development of the fully digital GSM protocols (also known as 2G to differentiate it from the first analog generation) primed the transition from analog to digital transmission. Current fourth-generation (4G LTE) wireless mobile telecommunications technology allows high-speed mobile ultra-broadband data transmission in bands extending to 2100 MHz. Specialized data modems known as *cell data radios* transmit and receive data using 3G protocols through the so-called EDGE-GSM system, where GSM is the 2G European telecommunications standard and EDGE is a backward-compatible 3G enhanced data-rate protocol employing the advanced *Eight Phase Shift Keying (8PSK)* phase shift modulation scheme. Eight 45° signal phase shifts are allowed, each carrying 3 data bits. High-speed internet protocols (3G, 4G LTE) allow command and control with visual display of the data output (Fig. 2.6) from remotely operated onsite computers as is used for HF radar installations.

5.2.3 Cable Connections for Coastal Ocean Observing

Cable connections of subsurface instruments to surface communications platforms in shallow waters are common. For buoy applications, instruments may be mounted within specialized steel cages physically incorporated into the buoy mooring system (Fig. 3.2). Power and communication cables are fairlead along the mooring cable to waterproof bulkhead connectors mounted on the payload bay housing.

Cabled shore-piercing systems are in use at some of the research installations discussed above such as LEO-15 off the New Jersey coast and the Martha's Vineyard Observatory off the eponymous island of the coast of Massachusetts. While they are extremely reliable and their operational costs are modest, initial costs of cable installation are largely prohibitive. Exposed cables crossing the shoreline are hazardous and subject to significant abrasion in the high energy (Pierce and Romanelli 1969) surf zone and are banned in many jurisdictions. An alternative, more environmentally sound installation entails directional drilling to tunnel beneath the shoreline zone, an effective but expensive technology common to the fossil fuel industry. Such installations normally require official permits from various government agencies, a lengthy process in most cases.

5.3 Acoustic Data Links for Coastal Ocean Observing

The seawater medium is largely opaque to electromagnetic radiation in the frequencies commonly used for radio communicationns (3 kHz to 300 GHz) with the exception of the extremely low frequency band (ELF 3–30 kHz). Naval submarines rely on this band for communications to depths of around 100 m and ranges of up to 10,000 km. However, the low frequency results in low data density allowing only brief communications such as a commmand to raise a conventional surface-piercing antenna to receive further information. Although some experimental progress in underwater radio transmission in the 1–5 MHz band has been reported (Lucas and Yip 2007), communication via electromagnetic means between submerged Tx/Rx antenna pairs to date remains commercially unavailable. Acoustic links constitute a viable (albeit slow) alternative to radio communications which, for reasons discussed above, are unviable in the subsurface ocean environment. The ocean is however an inherently noisy environment with acoustic signals ranging from 10 to 100 kHz and intensities of up to 100 dB from sources as diverse as earthquakes, explosions waves, wind, rain, shipping, ocean exploration, and the marine biota (Marsh 1969). Coastal underwater environments are particularly noisy due to the profusion of marine life and intense human activity. The challenges of acoustic communications in the noisy marine environment have been to some point overcome using digital wide band/spread spectrum communication schemes implementing multiple frequency shift keyed modulation or phase shift keyed modulation, data packaging, and multipath correction to achieve error-free data transmission. A recent breakthrough (Shi et al. 2017) adds an additional modulation dimension by impressing orbital angular momentum to the acoustic beam thus generating *helical acoustic vortex* beams.

Instruments for underwater data communication incorporating such features are known as *acoustic modems*. Bottom mounted acoustic wave and current meters and deep ocean tsunami monitors both communicate with surface buoys making use of acoustic modems as do deep-water moorings incorporating multiple instruments. Commercially available *acoustic modem* links, widely used in the offshore industry, operate at frequencies between about 10 and 50 kHz achieving ranges beyond 4000 m Sendra et al. (2016).

Sea bottom emplacements and ODAS buoys systems require periodic retrieval for instrument and equipment maintenance. Shallow water emplacements can be accessed by divers or buoy tender vessels but alternate means are necessary at greater depths. The *acoustic release* is an electronic instrument designed to allow recovery of buoys or sea bottom emplacements moored at depths beyond the reach of divers and buoy tenders. Their remote operation involves a complex mix of disposable ballast, and integrated buoyancy besides the acoustic release. Plastic encased spherical glass floats ensure the required buoyancy upon anchor release. The acoustic release acts as a structural component, essentially a shackle, but incorporates an acoustic transducer and circuitry tuned to a specific acoustic command sequence that prompts the opening of the mechanical component. In practice, the recovery vessel lowers a hydrophone into the water to deliver a coded signal unique to the instrument being addressed. To minimize malfunction contingencies, high-value packages may be secured by paired acoustic releases in a configuration such that release by either of the pair achieves release of the buoyed instrument package leaving behind only the disposable ballast. In general, shallow water instruments operate in the range of 20–50 kHz. Deeper deployments require lower frequencies to achieve better propagation in the seawater medium.

5.4 Satellite-Aided and Autonomous Underwater Navigation for Ocean Observing

Autonomous surface vehicles and other surface or surface-piercing vehicles such as gliders and buoys make use of the US *Global Positioning System* (GPS), a mainstay for earth surface location and tracking in many applications including ocean observing. The system consists of a satellite constellation orbiting at 20,200 km altitude in 6 orbital planes with 27 operational satellites and additional spare units in orbit. Each satellite carries multiple highly precise atomic clocks which allow a precise geographical fix computation upon interrogation of at least three satellites in direct line-of-sight by a surface platform. Data from additional satellites reduces the uncertainty perimeter and permits calculation of altitude above the geoid. Since the GPS satellites do not provide data relay, the autonomous platform must be provided with alternate telemetry solutions such as cellular networks and/or any of the satellite communications systems cited above. For accurate navigation, technical details for civilian GPS applications are available online at http://www.gps.gov/technical/ps/2008-SPS-performance-standard.pdf (US Department of Defense 2008).

The GPS system as noted previously is widely used for navigation by autonomous surface craft, gliders, and free drifting profiling buoys. Moreover, data buoys are commonly fitted with GPS receivers to confirm that the buoy remains within its designated *watch radius* and has not lost its mooring.

For mobile underwater vehicles, radio frequency-based remote navigation aids such as GPS are precluded because seawater is largely opaque to radio waves. *Inertial navigation*, accurate timekeeping plus knowledge of orientation and acceleration in three axes, provides a means for calculating vehicle trajectories in both time and space. Three-axis fluxgate magnetometer compasses are the mainstay for determination of instrument orientation within the earth's magnetic field. These instruments incorporate orthogonal copper windings around a magnetic core. Current induced in the windings depends on the orientation within the earth's magnetic field. Single or multiple three-axis array magnetometer assemblies thus provide accurate readings of the magnetic field. *Microelectromechanical devices* incorporate micro-machined cantilevered beams whose deformation provides the acceleration data. Modern inertial instruments incorporating multiple inertial accelerometers, tilt sensors, and vibrating gyroscopic sensors, usually mounted on a single board, constitute off-the-shelf *inertial measurement systems.*

Supplementing these measurements with acoustic means provides additional navigation and positioning data. *Doppler velocity logs* incorporate ADCP current measurements and acoustic bottom tracking as well as inertial navigation systems providing effective navigation at altitudes (distance from bottom) of up to 200 m. Bottom mounted acoustic beacons can provide additional reference for precise navigation within defined fields.

References

Khanna P. Microwave oscillators: the state of the technology. Microwave J. 2006. 7 pp. (Accessed 11/18/2017).

Lucas J, Yip CK. A determination of the propagation of electromagnetic waves through seawater. Int J Soc Underwater Technol. 2007;27(1):9 pp. ISSN 0141 0814.

Marsh HW. Underwater sound and instrumentation. In: Myers JJ, Holm CH, McAllister RF, editors. Handbook of ocean and underwater engineering. New York: McGraw Hill; 1969. pp. 3–4 to 3–30.

Nestlebush MJ. The geostationary operational environmental satellite data collection system, NOAA technical memorandum NESDIS, vol. 40. NOAA, Washington, D.C; 1994.

Pierce GA, Romanelli RP. Cable installation and repair. In: JJ Myers, CH Holm, RF McAllister (Eds.). Handbook of ocean and underwater engineering. Copyright by North American Rockwell Corporation. New York: McGraw Hill; 1969. pp. 3–4 to 3–30.

Sendra S, Lloret J, Jimenez JM, Parra L. Underwater acoustic modems. IEEE Sensors J. 2016;16:11.

Shi C, Dubois M, Wang Y, Zhang X. High-speed acoustic communication by multiplexing orbital angular momentum. Proc Natl Acad Sci U S A. 2017;114(28):7250–3. https://doi.org/10.1073/pnas.1704450114.

US Department of Defense. Global positioning system standard positioning service performance standard. 4th ed. Washington DC;.2008. https://www.gps.gov/technical/ps/2008-SPS-performance-standard.pdf. Accessed 4/30/2017

Chapter 6
Numerical Models for Operational Ocean Observing

Abstract Ocean general circulation models (OGCM) mathematically simulate ocean water mass movements making use of the hydrodynamic equations adhering to the conservation of mass, and energy through simplifying assumptions that allow operational algorithms. Models are built upon grids and computations are performed at grid nodes and propagated through the grid at predetermined time steps. Local coastal circulation models, addressing a smaller area, can implement denser grids providing greater detail. These however are commonly embedded or *nested* within large-scale OGCMs reducing computational demand by providing *boundary conditions* between the models. Operational *assimilation* of instrumental data constrains *model drift*, extending the fidelity of model forecasts. Lagrangian tracking of virtual particles released into or upon the water surface provides guidance for spill tracking and search and rescue operations. Spectral ocean wave models determine energy density across the wave spectrum allowing forecasts of wave heights, parameterized as significant or maximum wave height, wave period, and wave direction. Accurate coastal circulation and wave modeling requires detailed coastline and bottom topography databases as well as fine-grained model wind fields. Chemical models are being used to track and forecast ocean acidification and biological models are being tuned for prediction of harmful algal bloom occurrences.

Keywords General circulation models · Data assimilation · Structured grids · Unstructured grids · Coastal models · Wave models · Lagrangian tracking · Chemical models · Biological models

6.1 Constraints to Spatial and Temporal Resolution of Ocean Observing Models

Numerical models are synthetic digital representations of the physics, chemistry, and biology of the ocean. Three-dimensional physical models provide the basis for complementary optical, chemical, and biological models. Computations are

The original version of this chapter was revised. A correction to this chapter can be found at https://doi.org/10.1007/978-3-319-78352-9_9

J. E. Corredor, *Coastal Ocean Observing*,
https://doi.org/10.1007/978-3-319-78352-9_6

performed at spatial geographic grid nodes and propagated through the grid at predetermined time steps. Griffies and Treguier (2013) estimate a grid size on the order of 10^{27} nodes for the world ocean at the millimeter scale, and around 10^{10} time steps for a millennial simulation at 1 s resolution. Limitations in computational capacity thus necessitate constraint of spatial and temporal resolution. For global ocean models, spatial resolution on the horizontal is for historical reasons related to fractions of degrees of latitude that ranged from very coarse 2° resolution to the now common 1/12° resolution (equivalent to about 9.3 km). Nevertheless, as computational capacity increases, ultra-high resolution global implementations are now possible as is the case of NASA JPL's 1 km resolution sea surface temperature product (https://ourocean.jpl.nasa.gov/SST/. Accessed August 5 2017). Operational models are updated at periods of minutes to hours.

6.2 Physical Models for Operational Ocean Observing

6.2.1 Ocean General Circulation Models for Operational Ocean Observing

Ocean general circulation models (OGCM) constitute digital representations of ocean hydrodynamics at the global scale making use of the equations of fluid dynamics. Atmospheric forcing (pressure, wind); buoyancy differentials caused by solar heating, freezing, and melting; rain and runoff; and, in extreme, tectonic movements impart acceleration to fluid parcels setting up horizontal and vertical currents. Subsequent displacement and mixing of these water parcels is constrained by equations describing the conservation of mass, momentum, and thermal energy (the so-called *primitive equations*), by turbulence and by boundary interactions with the atmosphere, cryosphere, and ocean bottom. Since seawater density is parameterized to temperature, salinity, and pressure through the equation of state of seawater (IOC et al. 2010), a mass balance equation for salinity is incorporated. Modeled on a rotating earth, such representations incorporate planetary vorticity through the Coriolis parameter. A *free surface*, referred to as the height (H), is permitted at the air-sea interface (distinct from bathymetric depth (h) relative to the reference geoid) which allows modeling of the surface displacements associated to tides, eddies, waves, wave and wind-driven run-up, tsunamis, and other phenomena. Atmospheric forcing may be derived from global climatologies but is today more commonly derived from *atmospheric general circulation models* with and without assimilation schemes to incorporate meteorological or satellite wind observations. Various *truncation* schemes are applied to reduce computational demand such as the *hydrostatic approximation* that assumes hydrostatic equilibrium throughout the water column and the *Boussinesq* approximation which assumes constant density throughout the water column to simplify computation of horizontal momentum.

Ocean general circulation models are implemented on three-dimensional (3-D) spatial geographic computational grids. Models are thus categorized as to how *spatial discretization* is achieved for nodes upon which solution approximations to

Table 6.1 General ocean circulation models providing boundary conditions for high-resolution coastal models

Model	Vertical parameterization	Institutional support	Web site
Princeton Ocean Model (POM)	σ (terrain following)	Princeton University/ Stevens Institute of Technology	
Regional Ocean Model System (ROMS)	σ	Rutgers U./UCLA	https://www.myroms.org/
Hybrid Coordinate Ocean Model (HYCOM)	Hybrid; isopycnic in open ocean but z in mixed layer, transitioning to σ in shallow water	US National Ocean Partnership Program Consortium	https://hycom.org/
Nucleus for European Modelling of the Ocean (NEMO)	Hybrid: z/σ	European consortium	https://www.nemo-ocean.eu/

partial differential equations are computed. Many OGCM use the *finite difference* method on structured orthogonal curvilinear grids that conform well to large volumes in uniform basins. Numerical models advance the state of each node in the grid at successive or multiple interleaved *time-steps* (Williams 2009).

Sub-grid scale phenomena are not directly resolved through discretization. Since grid scales can be orders of magnitude greater than the scale of molecular diffusion and turbulent mixing, parameterization schemes are applied to specific phenomena ranging from molecular diffusion, through *salt fingering* (double diffusion of salt and heat), to turbulent shear, eddies, and internal wave effects. Appropriate parameter choice allows tuning of the model to specific basin features.

Models are further categorized as to the coordinate system chosen for grid computation following geopotential (z) or isopycnal (density (ρ)) surfaces, or assigning a constant number of depth-normalized *terrain-following* layers (σ). Each presents desirable properties that vary significantly in the ease of transitioning from the deep sea to the shallow water coastal setting. Geopotential grids may work well to simulate deep ocean conditions but are reduced to a few or only a single layer in shallow waters. Grids modeled along density surface may conversely simulate well variations across the main ocean thermocline in near surface waters but may lose detail in portraying water mass displacements along density surfaces in the deep ocean where density variations are minimal. *Hybrid* systems alternating among the above provide this flexibility. Among current OGCM models HYCOM, ROMS, POM, and NEMO serve coarse-resolution *boundary conditions* to finer resolution operational regional models (Table 6.1).

Data assimilation of near-real-time instrumental data is an essential feature of operational ocean models in use for coastal applications (Edwards et al. 2015). Computational assimilation of near-real-time data constrains numerical model data output rendering graphical interpretations such as maps, time series, profiles, and sections conforming to recent instrumental observations. In one computational scheme, statistical cost function procedures adjust model output (the forecast) in response to a coincident instrumental observation in order to produce a new output estimate (the analysis). Weights are ascribed to the data and prior model output such that the analysis output estimate minimizes the cost function (J). Thus the numerical estimate is smoothly and repeatedly nudged toward the observation. Optimal interpolation (OI) methods minimize cost function based on the inverse error co-variance matrix. Assimilation is complicated due to the limited number of observations in time and space and poor synopticity except in special cases such as satellite sea surface observations or HF radar. For these data-rich cases, analysis may be limited to known variability hotspots dispensing with analysis of low variability regions to optimize use of computational resources.

6.2.2 Coastal Ocean Hydrodynamic Models

Coastal users of hydrodynamic models, by and large, require finer-scale detail than that provided by OCGMs to explicitly model shallow water dynamics along convoluted coasts. Regional coastal ocean circulation models, at the most fundamental level, seek to resolve variations in sea level, currents, temperature, and salinity which variables encompass the bulk of physical variability of the coastal ocean. Such regional-scale models are commonly embedded or *nested* within an appropriate OGCM which provides the *boundary conditions* for the smaller, finer scale model. For a coastal model in an orthogonal framework, one open boundary (e.g., for an inner bay) can be specified. Oftentimes three open boundaries are specified encompassing an ocean domain along a single coastline. For an island or archipelago, the entire area may be within the nested model domain and thus four boundaries are specified. Exchange of mass, thermal energy, salt, and momentum across these boundaries (including free surface displacements) must be assimilated from the larger domain model. Enhanced viscosity and diffusivity *sponges* can be parameterized to a certain distance into and out from the nested model in order to dampen transfer across the boundaries (Edwards et al. 2015).

Regional hydrodynamic models now abound. Models based on structured grids such as ROMS in US coastal waters and NEMO in European waters are used extensively in regional implementations for operational coastal ocean modeling. Nevertheless, in recent years, *finite element* and *finite volume* methods that allow the use of unstructured topology representations, such as triangulation grids, have gained favor. These methods greatly facilitate construction of dense computational meshes along convoluted coastlines gradually transitioning to a sparser offshore mesh thus avoiding the large discretization errors incurred when applying structured grids. Examples of such models are the Finite Volume Community Ocean Model – FVCOM (Chen et al. 2003, 2006, 2007), the SELFE model (semi-implicit Eulerian-Lagrangian finite-element model for cross-scale ocean circulation) (Zhang and Baptista 2008), and the ADvanced CIRCulation Model – ADCIRC (see Xie et al. 2016 and references therein). ADCIRC features a versatile 2-D (nonstratified) version used extensively for inundation modeling (e.g., Xie et al. 2016). A detail of an unstructured grid developed for ADCIRC implementation in the CariCOOS region is shown in Fig. 6.1.

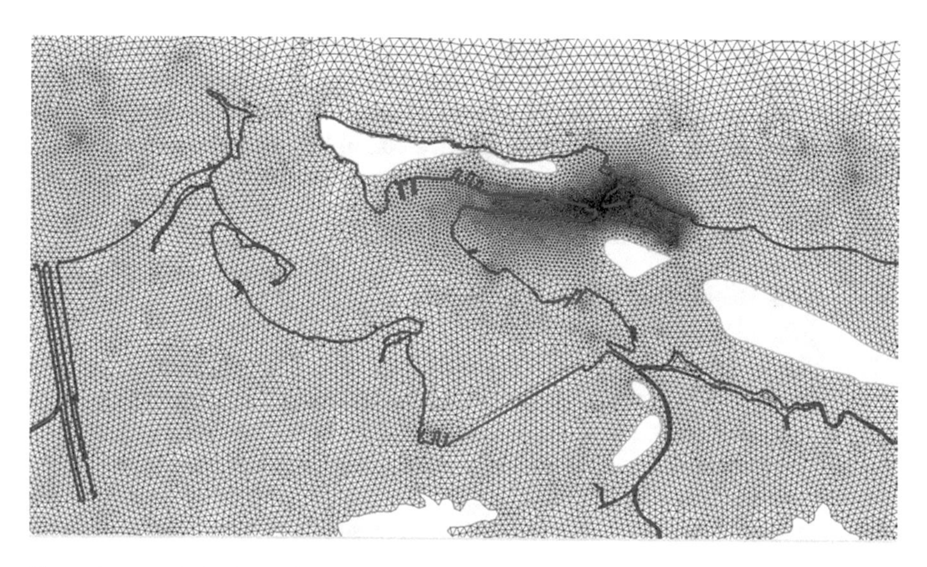

Fig. 6.1 Port of San Juan, Puerto Rico. Detail of the computational grid for the CariCOOS ADCIRC coastal inundation implementation

In the USA, NOAA supports a National Operational Coastal Modeling Program (NOCMP) with the mission of building and maintaining an Operational Nowcast and Forecast Hydrodynamic Model Systems (OFS) (https://tidesandcurrents.noaa.gov/models.html accessed 8/9/2017). These OFS supplement astronomical tidal predictions by incorporating additional free surface movements along the coastline taking into account storm tides and wave setup. OFS also provide hindcasts, nowcasts, and forecasts of current speed and direction, water temperature, and salinity. Specific forecasting OFS systems are targeted to cover critical estuaries, ports, and harbors, but diverse computational engines are in use for specific regions as described below.

Currently, operation regional NOAA OFS models for the US east coast include the Delaware and Chesapeake Bays OFS, using ROMS, the St. Johns River (Florida) Estuary, and the Ports of NY and NJ OFS, using the Environmental Fluid Dynamics Code, a model similar to POM in physics and computational code, the Northern Gulf of Mexico OFS operating a version of FVCOM and the Tampa Bay OFS using ROMS. Two high-resolution OFS are operational on the US West coast: the Columbia River Estuary OFS operating SELFE and the San Francisco Bay OFS using FVCOM. NOAA also supports 11 regional ICOOS systems operated by regional associations each of which either implements regional scale models to address region-specific problems or serves national scale models in region-specific formats.

An example of an institutional model is the New York Harbor Observing and Prediction System (NYHOPS), operated by Steven Institute of Technology in Hoboken, New Jersey. The model, based on the Princeton Ocean Model (POM), ingests meteorological and oceanographic data from a large number of sensors within the harbor, up the Hudson River and in the NY Bight, and provides forecasts. NYHOPS has received much attention due to its successful assistance in resolving a number of high-profile incidents in the urban estuary including the ditching of a commercial flight in the Hudson River in 2009 and urban inundation caused by Hurricane Sandy in 2012 (Blumberg et al. 2015). The urban inundation model, operating on a 3.1 m grid and using high-resolution LIDAR topography provides guidance for flood prevention intervention in preparation for future such events.

6.3 Coastal Ocean Wave Models

In contrast to hydrodynamic models which operate in a four-dimensional space/time frame (x, y, z, t), wave models operate in a five-dimensional space/time frame (x, y, t, θ, σ) where θ is the propagation angle and σ is the frequency (Booij et al. 1999; Mellor et al. 2008), a feature that increases the computational demand by one degree of freedom over that of hydrodynamic models. *Third-generation* wave models in use today incorporate varying wind fields (assimilated from satellite observations and/or operational wind models); resolve nonlinear wave-wave interactions, which are responsible for energy transfer across the wave spectrum; and explicitly model energy dissipation through bottom friction, bottom-induce wave breaking, and whitecapping. Such *spectral* wave models portray well the wave fields in the deep ocean. In nearshore waters where both waves and currents interact with convoluted topography and bathymetry model, performance depends greatly on the precision and resolution allowed by the digital representations of physical features (stored as *shapefiles* for geographic information system (GIS) software) and of the rugosity and consistency of these coastal features. These elements are crucial to accurate representation of wave evolution and coastal inundation. Bottom types are parameterized in terms of their frictional interactions with waves and currents through so-called Manning roughness coefficients (Benitez and Mercado 2015) or other empirical roughness parameterizations. Attenuation of wavefields with subgrid scale natural features such as floating sea ice must be similarly parameterized according to the myriad sea ice types. Accurate estimates for such parameterization are critical for successful modeling of wave dissipation in high rugosity environments such as grassy wetlands, the dense root structure of mangrove forests, and coral reefs (Lowe et al. 2005).

Experienced mariners are well aware that opposing wave and current fields make for choppy seas, a clear indication that accurate numerical representation of the wave field is incomplete without reference to ocean currents, especially nearshore where topography-induced jets occur. Such representation is achieved using fully coupled hydrodynamic and wave models requiring simultaneous runs of both models. As a result, it is only recently, as computational capacity has increased, that fully coupled models on unstructured grids have been achieved (Xie et al. 2016; Chen et al. 2013). These are largely run to provide forensic analysis of past storms or to develop storm surge atlases which are catalogs or look-up list for storm inundation preparedness. Inundation hazard databases are prepared carrying out multiple model runs with simulated hurricanes of different magnitude and size at different angles of approach and track offsets. Estimates of storm surge maximum of maximums (MOM) and maximum envelope of water (MEOW) are then derived for guidance in emergency management. The CariCOOS Storm Surge model, adapting a previous implementation of ADCIRC+SWAN (PADSWAN) to the northeastern Caribbean islands to support emergency response, risk assessment, coastal planning, and climate change analysis in Puerto Rico (Benitez and Mercado 2015), is an example of such storm surge atlases. Luettich et al. (2017) have also recently implemented the NOAA Sea Lake and Overland Surge from Hurricanes model (SLOSH) coupled to the SWAN deep wave model for the CariCOOS region providing an additional inundation and surge database.

Global operational wave models notably include the third-generation NOAA WaveWatch III™ (WWIII) implementation (Tolman et al. 2002) forced by NOAA's Global Data Assimilation System (GDAS) and the operational Medium-Range Forecast system (GFS). Numerous implementations of the Simulating Waves Nearshore (SWAN) model originally described by Booij et al. (1999) are also operational, many of these nested within WWIII. SWAN includes explicit formulations to incorporate the effects of currents on waves. NOAA uses such a scheme for its coastal Nearshore Wave Prediction System (NWPS) which incorporates RTOFS (NOAA's Global Real-Time Ocean Forecast System) currents.

The CariCOOS Nearshore Wave Model (Anselmi et al. 2012; see box) is an implementation of such a nesting scheme for Puerto Rico and the Virgin Islands. The model provides graphical time series forecasts for *virtual buoys* located throughout the coastal waters of PR and USVI some of these co-located with actual wave-measuring buoys.

The CARICOOS Nearshore Wave Model

The CARICOOS Nearshore Wave Model (Version 7.0 – last updated April 2016) is an operational wave forecast system based on the Simulating Waves Nearshore (SWAN) spectral wave model. It is forced by NOAA WWIII spectral boundary conditions and by the CARICOOS WRF 2-KM resolution operational wind model. It is based on a 1 KM resolution parent grid with 12 nested grids at a spatial resolution between 240 m and up to 10 m in some locations. The model is run twice per day (00Z and 12Z cycles) at the CARICOOS High-Performance Computing facilities. Forecasts are provided for 120 h on the parent grid and for 72 h in the nested high-resolution grids (http://www.caricoos.org/ accessed 12/1/2017).

Model performance was field validated using data output from a CariCOOS wave buoy located in Vieques Sound off the coast of Rincon, PR (northwest coast), and a CariCOOS ODAS buoy off Ponce PR (south coast). The nested model is shown to improve over the performance of WWIII alone (Anselmi et al. 2012). Model performance during the recent passage of Hurricane Maria across the islands, an extreme event, can be further judged through comparison with actual readings of a buoy moored in Vieques Sound to the east of the main island of Puerto Rico. Observations at the buoy over a 72 h period (September 19–21, 2017) during storm passage reported significant wave height (SWH) ascending to 6.09 m at 07:00 local time. The CariCOOS model SWH forecast of around 5.2 m underestimated the measured value by about 15% and anticipated peak SWH by about 2 h. Ample warning of significant wave action within the sound was thus provided by the 72 h forecast. Interestingly, while the supporting WRF wind model correctly anticipated the wind reversal later recorded by buoy anemometers, the wave model failed to capture the sudden drop and recovery of SWH following the initial peak seen in the buoy data coincident with observed wind reversal and 37 mb decrease in barometric pressure.

6.4 Lagrangian Tracking for Spill Response and Search and Rescue

Lagrangian particle tracking is a modeling application to extract Lagrangian trajectories from the Eulerian description provided by hydrodynamic models and/or HF radar output. Hindcast and forecast model runs provide estimates of trajectories following an initial event to the present and into the future thus facilitating search and rescue operations at sea and emergency spill response.

The US Coast Guard search and rescue operations (as well as those of many other coast guards worldwide) are planned using the Environmental Data Server (EDS), a data storage and retrieval system, as well as a software environment, the Search and Rescue Optimal Planning System (SAROPS). An advanced version of the Short Term Predictive System (STPS) (Barrick et al. 2012), a stochastic *Monte Carlo* model, now assimilates data from the IOOS mid-range HF radar network for SAROPS.

NOAA's Office of Response and Restoration has developed a *General NOAA Operational Modeling Environment* (GNOME) as a modeling tool to facilitate spill tracking and response as well as response training (Beegle-Krause 2001). GNOME operates in conjunction with hydrodynamic models which provide current speed and direction. A large number (>500–10,000) of virtual Lagrangian Elements (LE), also called splots (for spill dots), are released at the spill site and displacement vectors are computed for each time step. To portray stochastic dispersion, a user-supplied dispersion coefficient is applied so that diverging splots may be depicted as a widening cloud.

6.5 Chemical Models for Coastal Ocean Observing

Numerical chemical models incorporate equilibrium and kinetic rate constants plus mass balance formulations to resolve specific reactions such as those here described (Chap. 2) for inorganic carbon equilibrium. Adverse ecological consequences of ocean acidification have driven the development of fully coupled hydrodynamic/chemical models that allow examination of the effects on carbonate chemistry of wave forcing, sea level, geomorphology, rates of benthic metabolism and air-sea gas exchange. These models make use of empirical *piston velocity formulations* (estimates of the effective gas transfer rate) parameterized to wind velocity. The coral reef model proposed by Falter et al. (2013) is a one-dimensional example.

A more empirical approach is taken in the operational model implemented by NOAA's Coral Reef Watch Program (Gledhill et al. 2007) based on satellite and shipboard observations of SST and SSS to arrive at estimates of A_T and a first-order exponential parameterization of ΔpCO_2 to CO_2 solubility. These parameterizations allow solution of the inorganic carbon equations via CO2SYS. Surface maps of derived variables such as aragonite saturation index for the Caribbean region may be constructed and viewed through the use of ERDAP at http://cwcgom.aoml.noaa.gov/erddap/griddap/miamiacidification.graph (accessed 8/17/2017).

NOAA's GNOME oil spill model is complemented by a stand-alone oil weathering model, the Automated Data Inquiry for Oil Spills (ADIOS). ADIOS makes use of a database of the physicochemical characteristics of more than 1000 oil types to estimate temporal losses due to emulsification, evaporation, dispersion, and dissolution, and changes in physical properties such as density and viscosity to compute emulsion mass and spreading rates of the surface slicks.

6.6 Biological Models for Coastal Ocean Observing

Biological models seek to mathematically depict key processes in the biological food web. Deterministic models incorporate numerical formulations for physiological and trophodynamic variables. Photosynthetic rate is parameterized to light availability and temperature, while phyto- and bacterio-plankton nutrient uptake kinetics is parameterized to nutrient concentration through the *Michaelis-Menten hyperbolic expression*. Here, phytoplankton nutrient uptake rate varies linearly with concentration in the lower concentration range but conforms to a *saturation* asymptote at higher concentrations. Trophodynamic variables such as zooplankton grazing rates include similar *saturation* schemes and extend to higher trophic levels incorporating additional formulations for predator-prey interactions.

Applied biological models for coastal ocean observing have centered on the problem of harmful algal blooms (HAB) during which certain diatoms and dinoflagellates produce neurotoxins such as domoic acid or saxitoxin causing the phenomenon known as paralytic shellfish poisoning. These blooms cause severe economic disruption and public health concern. An experimental model for blooms of the dinoflagellate *Alexandrium fundyense* along the northeastern coast of the USA (He et al. 2008) incorporates coupled hydrodynamic (ROMS/ADCIRC) and population models are initiated with data from cyst surveys delivering surface maps of projected cell concentrations (https://products.coastalscience.noaa.gov/hab/gomforecast.aspx).

Empirical models are also used for operational implementation to address HAB occurrence. The California-Harmful Algal Risk Mapping Model (C-HARM), for example, uses the ROMS hydrodynamic model in conjunction with satellite ocean color data and a statistical model to determine the likelihood for occurrence of blooms of *Pseudonitzschia* (a diatom common to coastal California waters) that produces domoic acid, a neurotoxin that causes shellfish poisoning.

References

Anselmi C, Canals M, Morell J, Gonzalez J, Capella J, Mercado A. Development of an operational nearshore wave forecast system for Puerto Rico and the U.S. Virgin Islands. J Coast Res. 2012;28(5):1049–56.

Barrick D, Fernandez V, Ferrer MI, Whelan C, Breivik Ø. A short-term predictive system for surface currents from a rapidly deployed coastal HF radar network. Ocean Dyn. 2012;62:725–40. https://doi.org/10.1007/s10236-012-0521-0.

Beegle-Krause CJ. General NOAA oil modeling environment (GNOME): a new spill trajectory model. IOSC 2001 proceedings, Tampa, FL, March 26–29, 2001, vol. 2. St Louis: Mira Digital Publishing, Inc.;2001. pp. 865–71.

Benitez J, Mercado A. Storm surge modeling in Puerto Rico in support of emergency response, risk assessment, coastal planning and climate change analysis. 2015. http://coastalhazards.uprm.edu/downloads/Storm%20Surge%20Modeling%20in%20Puerto%20Rico%20in%20Support%20of%20Emergency%20Response-V10.pdf. Accessed 1/19/2018.

Blumberg AF, Georgas N, Yin L, Herrington TO, Orton PM. Street-scale modeling of storm surge inundation along the New Jersey, Hudson River Waterfront. J Atmos Ocean Technol. 2015;32:1486–97.

Booij N, Ris RC, Holthuijsen LH. A third-generation wave model for coastal regions 1. Model description and validation. J Geophys Res. 1999;104(C4):7649–66.

Chen C, Liu H, Beardsley RC. An unstructured, finite volume, three-dimensional, primitive equation ocean model: application to coastal ocean and estuaries. J Atmos Ocean Technol. 2003;20:159–86.

Chen C, Beardsley RC, Cowles G. An unstructured grid, finite-volume coastal ocean model (FVCOM) system. Oceanography. 2006;19(1):78–89. https://doi.org/10.5670/oceanog.2006.92.

Chen C, Huang H, Beardsley RC, Liu H, Xu Q, Cowles G. A finite volume numerical approach for coastal ocean circulation studies: Comparisons with finite difference models. J Geophys Res. 2007;112:C03018. https://doi.org/10.1029/2006JC003485.

Chen C, Beardsley RC, Luettich RA, Westerink JJ, Wang H, Perrie W, Toulany B. Extratropical storm inundation testbed: Intermodel comparisons in Scituate, Massachusetts J Geophys Res Oceans. 2013;118(10):5054–5073, https://doi.org/10.1002/jgrc.20397.

Edwards CA, Moore AM, Hoteit I, Cornuelle BD. Regional ocean data assimilation. Ann Rev Mar Sci. 2015;7:21–42. https://doi.org/10.1146/annurev-marine-010814-015821. Epub 2014 Aug 6.

Falter JL, Lowe RJ, Zhang Z, McCulloch M. Physical and biological controls on the carbonate chemistry of coral reef waters: effects of metabolism, wave forcing, sea level, and geomorphology. PLoS One. 2013;8(1):e53303. https://doi.org/10.1371/journal.pone.0053303.

Gledhill DK, Wanninkhof R, Millero FJ, Eakin M. Ocean acidification of the greater Caribbean region 1996–2006. J Geophys Res. 2007;113. https://doi.org/10.1029/2007JC004629. https://doi.org/10.1029/2007JC004629.

Griffies S, Treguier AM. Ocean circulation models and modeling. In: Fundamentals of ocean climate modelling at global and regional scales. Trieste: International Centre for Theoretical Physics; 2013. p. 2512–5. p. 57. http://indico.ictp.it/event/a12235/material/0/4.pdf.

He R, McGillicuddy DJ Jr, Keafer BA, Anderson D. Historic 2005 toxic bloom of Alexandrium fundyense in the western Gulf of Maine: 2. Coupled biophysical numerical modeling. J Geophys Res. 2008;113:C07040. https://doi.org/10.1029/2007JC004602.

IOC, SCOR and IAPSO. The international thermodynamic equation of seawater – 2010: calculation and use of thermodynamic properties. Intergovernmental oceanographic commission, manuals and guides no. 5. Paris: UNESCO (English); 2010. pp. 196.

Luettich E Jr, Wright LD, Nichols CR, Baltes R, Firedrichs MAM, Kurapov A, van der Westerhausen A, Fennel K, Howlett E. A test bed for coastal and ocean modelling. EOS. 2017;98(11):24–9.

Lowe RJ, Falter JL, Bandet MD, Pawlak G, Atkinson MJ, Monismith SG, Koseff JR. Spectral wave dissipation over a barrier reef. J Geophys Res. 2005;110:C04001. https://doi.org/10.1029/2004JC002711.

Mellor GLM, Donelan A, Oeya LY. Surface wave model for coupling with numerical ocean circulation models. J Atmos Ocean Tech. 2008;25:1785–807. https://doi.org/10.1175/2008JTECHO573.1.

Tolman HL, Balasubramaniyan B, Burroughs LD, Chalikov DV, Chao YY, Chen HS, Gerald V. Development and Implementation of Wind-Generated Ocean Surface Wave Models at NCEP. Weather Forecast. 2002;17:311–33. https://doi.org/10.1175/1520-0434(2002)017<0311:DAIOWG>2.0.CO;2.

Williams PD. A proposed modification to the Robert–Asselin time filter. Mon Weather Rev. 2009;137(8):2538–46. SSN: 0027-0644; eISSN: 1520-049.

Xie D-m, Zou Q-p, Cannon JW. Application of SWAN/ADCIRC to tide-surge and wave simulation in Gulf of Maine during Patriot's Day storm. Water Sci Eng. 2016;9(1):33–41.

Zhang Y, Baptista AM. SELFE: a semi-implicit Eulerian-Lagrangian finite-element model for cross-scale ocean circulation. Ocean Model. 2008;21(3–4):71–96.

Chapter 7
Coastal Ocean Observing Data Quality Assurance and Quality Control, Data Validation, Databases, and Data Presentation

Abstract High value of instrumental databases is attained and maintained through a rigorous quality assurance/quality control program involving care of the instrument and sensors to assure the quality of the data stream, careful instrument calibration, and continuous data quality control to ensure that data is flowing and to detect spikes and data gaps or instrument drift. In the case of remote sensing or numerical model databases, *vicarious calibration* or *field validation* is required. Instrument data and metadata are stored digitally using prescribed character codes and database formats that facilitate retrieval and visualization of selected data packets from large databases. Value of the data to the user is affirmed when knowledge is extracted from the data packet through visualization of data products that may include tables or time series data but most usually consist of a combination of graphics and text.

Keywords Quality assurance · Quality control · Data validation · Skill assessment · Databases · Data visualization

7.1 Introduction

Ocean observing data was still recorded by putting pencil to paper only a few decades ago. Most operational data is now recorded through electronic interrogation of a sensor and interpretation of the sensor state through calibration algorithms. Electronic sensors coupled to digital electronic processors vastly improve upon the performance of the previous generation of mechanical analog readout instruments in terms of data density and data quality. The glass/mercury thermometer, still in use in the second part of the twentieth century, is a case in point. These instruments were deployed by hand and read by eye allowing at best perhaps a few hundred raw data points per year per thermometer. Laborious calculations were further required to correct each thermometer reading. The data sampling rate of an electronic field instrument on the other hand is only governed in theory by the native response time

---

The original version of this chapter was revised. A correction to this chapter can be found at https://doi.org/10.1007/978-3-319-78352-9_9

J. E. Corredor, *Coastal Ocean Observing*,
https://doi.org/10.1007/978-3-319-78352-9_7

of the primary sensor. With a typical response time of 0.065 s such a sensor, routinely deployed to operate autonomously for up to 1 year, can collect over two million data points. In practice, data averaging and storage and transmission limitations constrain operational data yield of some sensor systems. Hourly binning of buoy surface data, for example, amounts to a modest 8760 data points per year but shorter sampling periods as required allow increased data rates. Many platforms perform vertical profiling as well thus increasing data rates by the number of vertical bins. Remote sensing systems providing synoptic coverage of 2-D fields yield orders of magnitudes greater data density. Aircraft or satellite-borne optical raster scanners and HF radar exemplify such remote sensing instruments. Additional large data sets are presently produced as numerical model output, an essential product of ocean observing. These large volumes of instrumental and virtual data are transmitted to data assembly stations, archived, rendered into value-added data products such as imagery and graphical output, and made available to end users.

7.2 Quality Assurance and Quality Control (QA/QC) for In Situ Ocean Observing Data

Extreme measures are taken to ensure the quality of instrumental data. Standardized *best practices* for instruments currently in use are codified in QA/QC manuals. The US IOOS has to date developed a series of twelve QA/QC manuals for a variety of sensors (https://ioos.noaa.gov/project/qartod/. Accessed 5/1/2018). It is the intent of IOOS to develop authoritative QA/QC guidelines for all core variables *addressing each variable as funding permits*.

Quality assurance refers to a number of active steps taken prior to, during, and after instrument deployment. These include, among others, appropriate instrument choice with the degree of resolution appropriate to the task at hand, rigorous instrument calibration, choice of instrument position aboard the platform free of interferences and abrasion, proper provision for error-free data transmission, steps for minimization of corrosion and biofouling, and post-deployment recalibration.

Numerical values of calibration coefficients obtained in the laboratory prior to sensor deployment allow conversion, for example, of the frequency output of a Wein-bridge oscillator coupled to a thermistor or conductivity bridge, to temperature in degrees Kelvin and practical salinity, respectively. Calibration must be traceable to an appropriate standard such as a certified platinum resistance thermometer for temperature or certified IAPSO standard seawater for salinity. It is common practice to return instruments to the manufacturer at prescribed periods for recalibration at the plant.

Instrument stability, and thus the need for recalibration, has been discussed previously. A few instruments such as ADCPs exhibit intrinsic stability allowing extended deployment (years), but others, notably chemical and bio-optical instrumentation subject to biofouling, are prone to rapid signal degradation requiring frequent cleaning and recalibration. IOOS manuals provide instructions for appropriate measures to reduce corrosion and biofouling allowing extended instrument operation.

In some cases, such as the Clark-type polarographic oxygen electrode, instrument drift is intrinsic to its operation. Redundant instrument deployment is always recommended to more readily detect sensor drift and to allow for failure of one component.

Profiling platforms pose special challenges to sensor systems due to abrupt gradients across physical, chemical, and biological discontinuities, thermoclines, haloclines, pycnoclines, chemoclines, and other gradients. Many instruments are thermosensitive such that their signal varies with temperature, requiring dedicated adjustment algorithms. Derived variables such as salinity (computed from temperature and electrical conductivity) are especially troublesome since *thermal mass* of the conductivity cell causes a lag in cell temperature relative to that sensed by the rapid response thermistor. Furthermore, the slower response time of the conductivity cell causes additional lag relative to the thermistor when traversing abrupt gradients. Manufacturers commonly provide software that can be programmed to carry out corrections to address these issues. Appropriate offsets depend on vertical vehicle velocity as well as relative sensor position on the vehicle and must be determined through testing.

Quality control consists of automated analysis of the data stream to detect flaws in transmission or instrument performance. IOOS protocols for quality assurance of real-time oceanographic data (QUARTOD) prescribe rigorous data testing prior to release. Thirteen standard tests are indicated for QUARTOD Q/C of T/S data (U.S. Integrated Ocean Observing System 2015a) provided as an example in the list below:

1. Time/gap test
2. Syntax test
3. Location test
4. Gross range test
5. Climatological test
6. Spike test
7. Rate of change test
8. Flat line test
9. Multivariate test
10. Attenuated signal test
11. Neighbor test
12. TS curve/space test
13. Density inversion test

Tests 1 and 2 address the quality of data transmission controlling for the arrival of the expected data package in the format required. Test 3 checks for proper georeferencing. Tests 4 and 5 ensure that the data received do not exceed the native instrument range or climatologically plausible values as determined by the operator. Tests 6 and 7 provide measures of proper sensor operation and require some form of user input with criteria based on climatological or other data. Test 8, like the eponymous movie, ensures that the sensor is not “dead.” Test 10 can serve to detect fouled or obstructed sensors. Test 11 relies on a comparable neighboring instrument but additional instrument-specific tests may be recommended. The above tests are widely applicable to many observing instruments. Tests 1 through 11 are standard for most instruments. Tests 12 and 13 are specific to T/S data pairs. A test for photic zone limit for radiance, irradiance, and PAR is recommended for optical instruments (U.S. Integrated Ocean Observing System 2015b), and a number of other tests referring to antenna performance are recommended for HF radar (U.S. Integrated Ocean Observing System 2016).

During Q/C testing, an additional data column is added to the array for inclusion of the standardized UNESCO suite of test flags as listed below:

Pass = 1	Data have passed critical real-time quality control tests and are deemed adequate for use as preliminary data.
Not evaluated = 2	Data have not been QC-tested, or the information on quality is not available.
Suspect or of high interest = 3	Data are considered to be either suspect or of high interest to data providers and users. They are flagged suspect to draw further attention to them by operators.
Fail = 4	Data are considered to have failed one or more critical real-time QC checks. If they are disseminated at all, it should be readily apparent that they are not of acceptable quality.
Missing data = 9	Data are missing; used as a placeholder.

7.3 Experimental Validation of Remote Sensing and Ocean Model Output Data

A model starts to have skill when the observational and predictive uncertainty halos overlap (http://www.meece.eu/documents/deliverables/WP2/D2.7.pdf Accessed 8/10/2017).

Since observational gaps do not allow for continuous synoptic real-time sampling in time and space, and model output is similarly limited by computational capacity, both views provide only approximations to the instantaneous state of sampled (or derived) variables. Moreover, inherent instrumental and computational errors occur in both quasi-synoptic remote sensing and model output. Experimental validation known as *vicarious calibration* in the case of remote sensing or as *skill assessment* in numerical model forecasting is essential to objective assessment of instrumental data quality and forecast accuracy. Properly designed validation experiments allow the observer to adjust instrument response and the modeler to identify and correct model deficiencies.

Assessment of instrument response can sometimes only be achieved via indirect calibration. Substantial resources for example are expended in calibration of variables such as near-surface Chl *a* and CDOM derived from spectral reflectance data

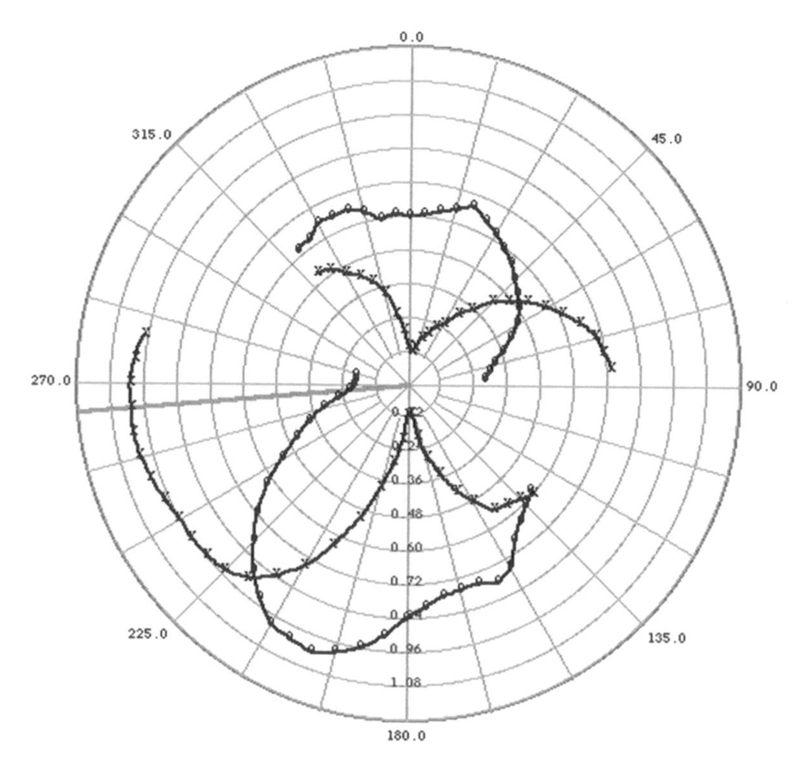

Fig. 7.1 Measured Rx antenna pattern denoting reception anomalies relative to the ideal pattern due to electromagnetic interference in the vicinity of the antenna

from satellite-borne radiometers. Once deployed, performance of instruments aboard these craft degrades requiring a sustained calibration effort throughout the useful instrument life. Brown et al. (2007) describe a buoy system dedicated exclusively to vicarious calibration of satellite ocean color measurements.

Surface ocean current measurements using HFR depend in large part on established physics of deep ocean wave propagation requiring little calibration. Nevertheless, two factors contribute to degrade signal quality: decreasing wave heights result in less data return increasing statistical uncertainty and electromagnetic interference from ferromagnetic bodies in the vicinity of the Rx antenna distorts the ideal pattern of orthogonal *figure 8* sensitivity lobes of the paired coils. Data correction is achieved through direct experimental measurement of the true antenna pattern, and data validation is achieved using Lagrangian drifters. Initial Rx antenna reception patterns are assumed to reproduce an ideal pattern of double crossed eights such as a four-leaf clover. Actual antenna reception patterns are subject to reception anomalies induced by ferromagnetic materials or electronic equipment in the vicinity of the antennae emplacement (Fig 7.1).

Antenna reception patterns are initially and periodically thereafter determined by deploying custom designed transponders coupled to GPS receivers in circular sections centered on the antenna. Coupled transponder, GPS, and antennae and are physically walked along the beach around the Rx antenna or deployed aboard a small vessel or aerial drone. Periodic antenna pattern tests are recommended especially if electromagnetic anomalies in the vicinity of the antenna are variable. Ships arriving and departing from port are an example. Measured patterns improve accuracy as objectively determined through graphical superposition and root mean square statistical analysis (Corredor et al. 2011) when field validated with Eulerian current meters and/or tidal current predictions.

Model validation may be performed comparing the performance of different models such as the tests employed by Chen et al. (2007) demonstrating superior performance of FVCOM implemented on an unstructured grid compared to POM and its derivative ECOM-si implemented on the structured curvilinear coordinate transformations. Haldivogel et al. (2008) apply Willmott's index of agreement (one to zero) to ROMS simulations of the tidal estuary of the Hudson River paired to instrumental data from bottom mounts and tight wire mooring arrays and hydrographic surveys along the estuary achieving average indices about 0.8. For wave models, data from wave measuring buoys within the data domain are often used. Anselmi-Molina et al. (2012) use RMSE and Willmott's index of agreement to validate the operational CariCOOS implementation of SWAN in nearshore waters of Puerto Rico and the Virgin Islands with reference to buoy data.

Willmott and collaborator's recent criticism of the use of RMSE (Wilmott et al. 2017 and references therein) bears consideration since they argue that *squaring each error often alters – sometimes substantially – the relative influence of individual errors on the error total.*

7.4 Ocean Observing Databases

Observing systems store data at local servers and submit copies to central storage facilities adhering to strict formatting, query, and delivery protocols. Distributed sensor network data sets are accompanied by information known as the *metadata*, descriptors of the data itself. Primary data is digitized to binary code but such ancillary metadata information includes descriptive text of temporal and spatial data such as date, time, latitude, longitude, and depth as well as instrument information such as calibration coefficients. *The UNICODE consortium*, a not-for-profit corporation, offers a unique means of identifying non-numeric text through *UNICODE*, a set of unique digital descriptors of language characters and symbols. One well-known character subset supported by *UNICODE* is the *American Standard Code for Information Interchange* (*ASCII*) which encodes the first 128 *UNICODE* characters. Computer keyboards in English and other Latin script languages are encoded in *ASCII*. Most other worldwide character sets (e.g., Cyrillic or Arabic) are further encoded within *UNICODE* as well as a large number of grammar, syntax, and typesetting instructions.

Early electronic *relational databases* constitute a simple means of storage and retrieval whereby data is stored in flat files, in essence 2-D data arrays addressable through the Structured Query Language (SQL). Data transfer was executed through the well-known 1-D *comma separated values* files characterized by the *.csv* extension. Modern multidimensional matrices provide for more efficient data storage and more versatile file transfer schemes. Formats in current use include the extensible markup language (XML) and geography markup language (GML), based on XML grammar. A highly capable format, the *network common data format* (netCDF), originally developed for atmospheric and meteorological science, is available from *UNIDATA*, a not-for-profit consortium of the University Corporation for Atmospheric Research (UCAR) with US National Science Foundation support. This simple machine-independent interface allows storage and access of n-dimensional data arrays together with their *metadata* through a set of functions supported for programming languages such as C++ and FORTRAN as well as for operating systems such as UNIX and Windows.

A number of *Data Access Protocols* (DAP) further facilitate data distribution and use through the internet and allow *data discoverability* across multiple computing platforms. The *Open-source Project for a Network Data Access Protocol* (sustained by the nonprofit OPenDAP organization) has received wide acceptance for oceanographic data retrieval. Thus, while netCDF files are available through the *File Transfer Protocol* (FTP), they may be more expeditiously acquired through the common *HyperText Transfer Protocol* (HTTP) using OPenDAP. UNIDATA, a community of US education and research institutions supports the *Thematic Real-time Environmental Distributed Data Services* (THREDDS). THREDDS data servers facilitate data discovery through detailed catalogs of the data sets available. Accessing a THREDDS server will allow discovery of the data subset required and will provide several options for data download including OpenDAP. Reliable data arising from a distributed sensor network is effectively served to scientists and advanced users through the above methods allowing use of the investigators operating systems and software package of choice such as IDL, JAVA, MATLAB, ORIGIN, or EXCEL.

In the USA and within the EU, data retrieval is further facilitated through the *Environmental Research Division's Data Access Program* (ERDDAP) incorporating all the formats and protocols discussed above providing a *Unified Access Framework*. Massive environmental data sets from satellite sensors and numerical model output as well data from in-water assets stored in diverse data servers are not only available for download but also for graphic display in a number of formats ranging from area maps to vertical sections and time series.

7.5 Ocean Observing Data Visualization for Environmental Awareness

The primary mission of an institutionalized ocean observing system is to render data arising from a local distributed sensor network, from satellite data, and from numerical model forecast output in forms that heighten the environmental awareness of stakeholders to support their safe, effective, and environmentally sound engagement in ocean-related activities. In the USA, IOOS has expended significant efforts to assess such needs. Stakeholders and experts have identified a set of 26 core variables of interest (see http://www.iooc.us/ocean-observations/variables/) including among others: tides, winds, waves, currents, seawater temperature, seawater salinity, nutrients, dissolved oxygen, and ocean color. Nevertheless, certain data sets are particularly sought in specific regions. The immediate negative impact of ocean acidification in the Pacific Northwest points to seawater pH as a variable of increasing significance to commercial interest in that region. Another example is the documentation of fog occurrence which is of importance to commercial and fishing interests in the Gulf of Maine and the US Pacific Northwest coast but is of little interest along Gulf of Mexico or the southern coast of California, or for stakeholders in tropical island regions. Similarly, tidal amplitude is most of the time of little concern in microtidal regions such as the northeastern Caribbean with a range of less than 1 m, but of permanent concern in regions such as the Gulf of Maine where it may reach above 8 m.

The challenge to the ocean observing organization, once stakeholder data needs are determined, is to deliver data or graphics that effectively serve stakeholder needs. For web page display, many organizations offer a surface map depicting regional observing assets in the water (including buoys, gliders, ocean bottom emplacements), or on the coast (including HF radar and weather stations). These assets are depicted as addressable icons which, upon interrogation, lead to dedicated pages containing data tables and graphs. Typically, the latest data from the suit of instrument aboard the platform of interest are provided as a single numerical readout and additionally as a self-refreshing table, as analog readout virtual dials, or as self-refreshing bar and line graphs. Timeline graphs for pertinent values may be brought up for varying time periods exposing periodic variability such as daily temperature fluctuations or yearly salinity excursions. Perusal of such figures allows a rapid visual assessment of prevailing maxima and minima providing a framework for enhanced understanding of existing conditions. More specialized renditions incorporate an additional patterned or color-coded third dimension to geographic or time series data. Synoptic georeferenced satellite imagery that superposes color-coded temperature, salinity, chlorophyll, turbidity, or other surface properties upon two-dimensional latitude/longitude is the most readily recognizable. Time series data at different depths may likewise be plotted using the horizontal axis as time and the vertical as depth (commonly referred to as **z** in oceanographic parlance). Data from autonomous gliders and towed undulating bodies can also be plotted in this fashion using the horizontal axis to depict distance along the vehicle track or time since deployment.

Numerical model data output is often depicted as color-coded maps. Model output provides forecast guidance, but caveats are normally attached to such graphics to address liability issues. Timeline graphics for in situ stations incorporating both instrument data and numerical model output data are instructive in allowing rapid evaluation of model performance. Little reliance can be placed on model forecast projections if hindcast data do not match up well to instrument output.

Applets for mobile phones and tablets incorporating these approaches further extend the immediacy of data access to users at sea, even aboard small craft. Today it is possible for most, anyone with access to a *smartphone*, a cellular telephone with internet capability, to be aware of the state of wind or current speed, of temperature, and of salinity (and of most of the other variables discussed in this book) at any point throughout a distributed observing network in near-real time depending only on the programmed reporting lag. ODAS buoys and HF radar stations for example typically report once an hour, while gliders may exhibit a time lag of three or more hours between surfacing. *Virtual buoys*, fixed-location output of data-ingesting numerical models, provide nowcasts and forecasts in digital and graphical format for a wide number of fixed sites at sea, many of these user selected. Representative graphical products from the 11 US ICOOS regional observing systems may be viewed through access to the IOOS home page at: https://ioos.noaa.gov/ (accessed 1/20/2018).

References

Anselmi-Molina CM, Canals M, Morell J, Gonzalez J, Capella J, Mercado A. Development of an operational nearshore wave forecast system for Puerto Rico and the U.S. Virgin Islands. J Coast Res. 2012;28(5):1049–56.

Brown SW, Flora SJ, Feinholz ME, Yarbrough MA, Houlihan T, Peters D, Kim YS, Mueller J, Johnson BC, Clark DK. The Marine Optical BuoY (MOBY) radiometric calibration and uncertainty budget for ocean color satellite sensor vicarious calibration Proceedings of the SPIE optics and photonics; sensors, systems, and next-generation satellites XI. 2007;6744:67441M.

Chen C, Huang H, Beardsley RC, Liu H, Xu Q, Cowles G. A finite volume numerical approach for coastal ocean circulation studies: Comparisons with finite difference models. J Geophys Res. 2007;112(C3):C03018.

Corredor JE, Amador A, Canals M, Rivera S, Capella JE, Morell JM, Glenn S, Handel E, Rivera E, Roarty H. Optimizing and validating high frequency radar surface current measurements in the Mona passage. Mar Technol Soc J. 2011;45(3):49–58.

Haldivogel DB, et al. Ocean forecasting in terrain-following coordinates: formulation and skill assessment of the Regional Ocean Modeling System. J Comput Phys. 2008;227(7):3595–624.

U.S. Integrated Ocean Observing System. Manual for real-time quality control of in-situ temperature and salinity data version 2.0: a guide to quality control and quality assurance of in-situ temperature and salinity observations. Silver Spring: NOAA US National Atmospheric and Oceanic Administration; 2015a. p. 56.

U.S. Integrated Ocean Observing System. Manual for real-time quality control of ocean optics data: a guide to quality control and quality assurance of coastal and oceanic optics observations. Sivler Spring: NOAA, US National Atmospheric and Oceanic Administration; 2015b. p. 46.

U.S. Integrated Ocean Observing System. Manual for real-time quality control of high frequency radar surface currents data: a guide to quality control and quality assurance of high frequency radar surface currents data observations. Silver Spring: NOAA; 2016. p. 58.

Willmott CJ, Robeson SM, Matsuura K. Climate and other models may be more accurate than reported. EOS. 2017;98(9):13–4.

Chapter 8
Planning, Implementation, and Operation of Coastal Ocean Observing Systems

Abstract Operational integrated coastal ocean observing systems (ICOOS) serve large stakeholder groups whose needs and expectations must be carefully assessed in order to design systems that provide responsive and useful data products. System design includes the selection of sensor suites and platforms appropriate to stakeholder needs, informed selection of emplacement sites such as to provide appropriately representative data for the area under study, and design and implementation of easily understandable and informative data products. Data-assimilating numerical models are custom designed for areas of interest to provide forecasts applicable for hours to days. Fixed platform emplacement may entail rigorous permitting processes that may extend over months to years. Fixed platform emplacement, recovery, and redeployment may also require heavy equipment and appropriate vessels as well as divers. Mobile platforms must be deployed and recovered from capable platforms, usually manned vessels. Sensor instruments mounted on fixed or mobile platforms require periodic maintenance to assure data quality. Data products must be served in as wide a variety of media as possible for effective and timely delivery to the stakeholder.

Keywords Stakeholder input · System design · Regulatory permitting · Platform emplacement · Command and control · Data products · System maintenance

8.1 Ocean Observing Data Needs Assessment

Coastal ocean observing is a costly endeavor requiring expensive sensors and platforms and the efforts of large numbers of personnel trained in a myriad of fields including those engaged in purely administrative duties, those involved in data handling, numerical modeling and digital product development, and the sea-going scientists, technicians, and students who actually emplace and service the necessary equipment. Legal counsel may be required in drafting bylaws for incorporation. The

---

The original version of this chapter was revised. A correction to this chapter can be found at https://doi.org/10.1007/978-3-319-78352-9_9

J. E. Corredor, *Coastal Ocean Observing*,
https://doi.org/10.1007/978-3-319-78352-9_8

effectiveness of these efforts is highlighted by the role these systems have played in a few catastrophic incidents such as hurricanes and large-scale accidents. More difficult to gauge are the economic benefits brought about through the facilitation of safe and effective activity in the coastal marine environment and lives saved.

Assessment of ocean observing data needs of stakeholders operating in the coastal ocean should be undertaken early-on using a wide variety of means of enquiry in order to develop a detailed picture of user requirements. Identifying the stakeholders is perhaps the most challenging initial undertaking. Personal visits to state and provincial government authorities and to maritime business enterprises, recreational operators, and travel agents focusing on the marine sector is an initial step. Such visits may lead to the identification of additional stakeholders and open the doors to addressing more structured forums such as operator associations, scientific societies, environmental groups, and government-promoted user associations. Among the latter, regional Harbor Safety and Security Committees, promoted by the Unites States Coast Guard, constitute welcoming fora for planning and promotion of ocean observing systems. Scientific societies, especially those focusing on ocean science, hold well attended yearly and biannual meetings and symposia where sessions dedicated to ocean observing are now quite common and additional stakeholders may be identified.

Once a representative user database is developed, internet-based polling services can provide effective and economical means for refining data needs assessment. Expert discussion of available assets allows assessment of data gaps and observation voids. Collective exercises (workshops, forums, and invited lectures) serve to contrast existing observing products to user needs and thus to the design of observing systems to fill identified data gaps. Prioritization exercises can help to decide upon the urgency of data needs and balance these against budgetary constraints.

A CariCOOS Stakeholder

In US jurisdictions, protocols are in place for the posting of weather-related port security restrictions based on official forecasts emitted by the National Weather Service. The decision as to when to open the port is left to the Port Captain. A former Port Captain at San Juan, Puerto Rico, a US Coast Guard officer, summed up these intangibles well remarking that, prior to the emplacement of a data buoy near the mouth of San Juan Bay, he was obliged to send out a small boat to gauge whether sea state was favorable enough to open the port for normal operations following the passage of heavy weather events, a hazardous and time-consuming operation. After buoy emplacement, objective numerical data on sea state has become available within the hour relieving the economic consequences of port closure to maritime interests while ensuring crew, ship, and port safety.

8.2 Planning Coastal Ocean Observing Systems

8.2.1 Instrument and Platform Selection

With data needs established and prioritized, an observing system may be designed to meet these needs. Data needs and budgetary constraints established during prioritization exercises will help to select an appropriate and affordable suite of instruments and platforms for autonomous deployment and numerical models for implementation.

Instruments most commonly specified for autonomous deployment center on measurements of tides, temperature, salinity, wave height, period and direction, current speed and direction, wind speed and direction, optical properties, dissolved oxygen, CO_2 partial pressure, pH, nutrients, hydrocarbons, chlorophyll a (Chl a), and colored dissolved organic matter (CDOM). More sophisticated laboratory analysis of phytoplankton samples may be important in regions prone to red tide occurrence.

Price, performance, reliability, and availability of instruments and platforms may vary widely and care must be taken in choosing the most appropriate. Shoddy manufacture can result in poor data quality and significant down-time with the consequent data gaps. Care should thus be taken in choosing manufacturers and dealers. Background checks of all suppliers should be undertaken prior to any purchase. Reputable manufacturers and dealers will normally provide the prospective customer with a list of previous customers whom the buyer is free to consult regarding equipment performance, promptness of delivery, and refurbishing and/or recalibration services and general responsiveness to user needs. Satisfied customers will generally be voluble in their praise, while unsatisfied customers might be more reticent. While equipment manufactured and marketed by foreign firms may, at first impression, prove attractive in terms of price and/or performance, procurement, refurbishment, and repair of such equipment may face high shipping charges and substantial hidden costs due to delays in shipping and service responding to foreign and local customs procedures. The user of such equipment may face the need to purchase backup units to fill these voids. However, even if the operator counts on responsible suppliers, equipment redundancy is usually a necessity. ODAS buoys, for example, are typically recovered and refurbished once a year. Equipment such as temperature/salinity units normally require shipment to the manufacturer for refurbishment and recalibration. Turn-around time for these procedures (including shipping time) is on the order of weeks to months while ODAS buoy recycling, from recovery to redeployment can be, barring unforeseen circumstances, a matter of days to weeks. New or properly refurbished instruments must thus be on hand so as not to delay redeployment. Standard practice is consequently to have a second suite of instruments available for immediate installation on all buoys.

Redundant, overlapping, and complementary platform/instrument combinations will provide for a more robust observing system. For surface ocean current measurements for example, availability of both a HF radar system and in situ current meters will allow mutual validation during operation of both systems and a fallback capability in case of failure of one or the other. Similarly, surface wave-powered vehicles or submersible glider transects can be planned so as to traverse the vicinity of fixed assets for data validation. Data from existing expeditionary serial observing efforts can profitably be incorporated into the observing system's products and autonomous vehicles can similarly be programmed to traverse such observing stations.

8.2.2 Platform Site Selection and Regulatory Constraints

Single-point measurements at dock sites or on cabled nearshore pilings will prove to be the least expensive solutions for mounting and operating autonomous instruments. Ocean Data Acquisition (ODAS) buoys may provide high-quality surface and vertical profile data. However, such systems are expensive to operate and maintain so their number will necessarily be limited. System siting, to be as representative of local conditions as possible, is consequently of utmost importance. Prior knowledge of local water mass movements and water residence times can aid significantly in choosing appropriate deployment sites for ODAS buoys. Knowledge of the nature and number of stakeholders to be served by a particular ODAS buoy will serve to further prioritize deployment sites. Likewise, as discussed previously, knowledge of bottom community composition will aid in selecting sites where environmental impact is minimized.

Site Selection Strategies for ODAS Buoys in Puerto Rico and the US Virgin Islands

ODAS buoy site selection in the Puerto Rico (PR) and the US Virgin Islands (USVI) archipelago is instructive. These Islands share a common insular platform formed by the Antillean Arc. The islands face the open western tropical Atlantic Ocean (WTA) to the north and the Caribbean Sea to the south. Ocean conditions can be dramatically different in these two basins. Whereas the WTA basin is characterized by very clear oligotrophic surface waters, a deep pycnocline (reaching down to 100 m and beyond) and powerful long period winter swells, Caribbean surface waters are for most of the year, under the influence of Amazon and Orinoco River waters and thus exhibit shallow pycnoclines (tens of meters), higher CDOM content, and more abundant phytoplankton. The Caribbean surface wave spectrum, largely conditioned by trade wind forcing, is dominated by moderate wave heights and shorter periods save for the occasional hurricane when monstrous waves may occur. Accordingly, in planning the CariCOOS network, initial ODAS buoys were deployed on either of these coasts of Puerto Rico to provide representative

(continued)

(continued)

data for both basins. Additional ODAS buoys were then emplaced to the south of St. John, USVI, and in Vieques Sound between PR and the USVI. Additionally, a specialized wave buoy was positioned off the west coast of PR at Rincon, a popular surfing spot where large swells are common.

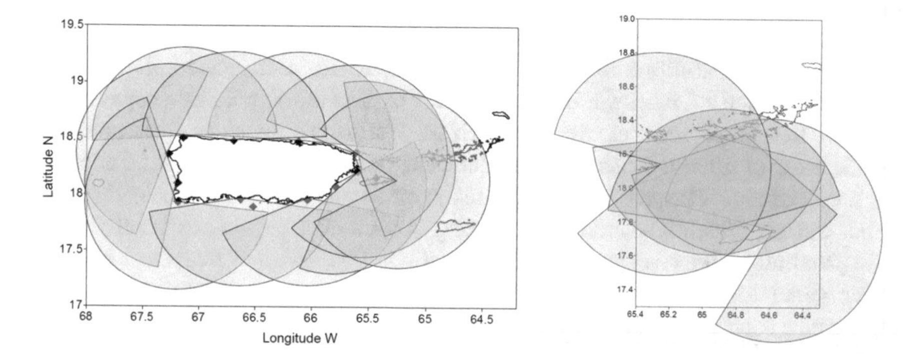

Fig. 8.1 Prototype mid-range HF surface current radar network for coastal waters off the island of Puerto Rico (PR) and the US Virgin Islands. Nominal range of 80 km is assigned to the individual stations

Emplacement of HF radar networks operating directional antennae (common equipment throughout the USA and elsewhere) requires significant forethought and planning since *vector solutions* (an explicit numerical expression of the surface current in terms of latitude, longitude, vector bearing, and vector magnitude) requires the input of two or more *radial* coincident vectors pointing toward or away from individual observation posts (antenna sites). Solutions are strongest in the antenna mid-range and at data points where radial observation pairs are orthogonal (90°) to one another. Solutions weaken as the angle between signals widens or contracts becoming indeterminate at small angles and approaching 180° (see Sect. 2.2.3 for a detailed description of HF radar operation). The planning effort requires maximizing area coverage while avoiding data gaps. Proper positioning maximizes far-field cover while optimizing field-of-view overlap to minimize the data gaps in the nearshore between observation emplacements, where solutions are weakest. For this reason, observation posts along linear coastlines are spaced at distances appropriate to antenna range varying from about 40 km for short-range units to over 100 km for long-range units. Bays and semi-enclosed water bodies offer the possibility of multiple antennae coverage. Isolated capes may however offer unobstructed fields of view with effective coverage exceeding 180°. Emplacement on islands within antenna range can greatly aid in increasing coverage while minimizing the number of antenna emplacements. Examples of both are provided in Fig. 8.1.

HF radar emplacements require electrical power for effective Tx antennae output and to operate control modules, Rx units, computers, radio or cell data Tx/Rx communication, and climate control units as required. While these requirements are moderate (about 3–5 kW per observation post), they are indispensable. Most units currently installed make use of gridline power thus foregoing many favorable deployment sites for lack of alternate power sources. A prototype Remote Power Module (Statsewich et al. 2011), which offers the prospect of true autonomy in HF radar emplacement, is described in Sect. 4.1 of this book. Self-powered, helicoptered and trailered autonomous HFR systems for rapid deployment have been successfully operated in western Florida and Norway (Whelan et al. 2010).

As power availability allows, HF radar antennae are best located close to the land-sea interface at low elevation to effectively exploit the conductive properties of seawater that allow sensing over the horizon. Unfortunately, since the nearshore zone is a site of intense human activity, ferromagnetic and electromagnetic anomalies are common. Massive steel structures such as ship hulls and dockside steel works can significantly distort the resulting Rx antenna pattern, especially if in close proximity. Likewise, large electromagnetic fields created by nearshore electric power generators may significantly degrade antenna performance. Intermittent interference such as that caused by ship arrivals and departures, by power generator output variability, or by alternating power sources in close vicinity to HF radar emplacements can confound interpretation rendering the data less accurate. Such sites will report large errors which even after subject to statistical analysis can result in uncertainty envelopes beyond operational usefulness. Siting alternatives should then be explored.

Government agencies with jurisdiction over the coastal and nearshore zones, such as those regulating natural resources, coastal navigation, and environmental health, may require permits for fixed platform emplacement. Detailed proposals may be required describing the platform, platform emplacement and recovery procedures, possible environmental impact of the proposed emplacement, and the hazards to navigation that the platform may pose.

Deployment of oceanographic and meteorological instrumentation on existing private or government owned docks is usually straightforward requiring only radio emission permits if contemplated. Deployment aboard existing aids-to-navigation (pilings and buoys) requires extensive negotiation with the operator, usually the local coast guard or equivalent body. Equipment redesign may be necessary to meet regulatory requirements.

For moored emplacements, standard practice is to select several alternate sites and to rank these according to desirability weighted against environmental impact and hazard to navigation. Once a site is approved, regulatory agencies may prescribe a moratorium to actual deployment while the emplacement is charted to warn mariners of its existence. Legal assurances must in any case be procured to cover liability in case of damage caused to third parties. Moored buoys inadvertently set adrift may, for instance, pose a particular hazard to navigation that must be anticipated.

While HF radar antennae emplacement is preferably close to the seashore, emplacement on adjoining private or public property may ease the regulatory permit load. A provisional blanket license may be issued for commercial directional HF radio emission but interference with other higher priority signals can precipitate revocation.

8.2.3 Selection, Sub-sampling, and Optimization of Satellite Imagery

Satellite imagery allows the synoptic visualization of dynamic events such as the onset, course and abatement of upwelling events, river plume dispersal, red tide occurrence, and sea ice dynamics at the coastal scale. Moreover, the satellite data underpinning these graphical representations is assimilated by numerical models allowing increased confidence in model forecasts. Access to recurrent operational satellite data for sea surface temperature, ocean color, and derived products such as near surface Chl a, CDOM custom in the ICOOS area of influence is normally forthcoming from NASA, NOAA, EUMETSAT, their Indian and Japanese counterparts and commercial operators, but significant data processing is usually necessary to develop graphical products for the local scale responsive to the full range of variability of regional features.

The geographical limits of the region of interest must be established so that code may be written to extract the pertinent data subset. Likewise, the regional variability range of these features (temperature, or Chl a, for example) must be established in order to constrain the color pallet to the regional range. In tropical regions, for example, a temperature color pallet ranging below about 20 °C will result in a featureless depiction and conversely so if high temperature brackets are allowed in temperate or arctic regions.

8.2.4 Selection and Design of Numerical Model Implementations

Choice of numerical models for incorporation of model output into the data suite served by an observing system will be informed by data needs ascertained during the stakeholder engagement process. While the basic model code will be easily available, several preliminary steps must be taken prior to model implementation. Local bathymetry is a powerful constraint to model domain delimitation since the boundary between the regional model and the parent global circulation model (GCM) should extend beyond the shelf edge into deep water so as to avoid spurious effects due to reduced performance of the GCM in shallow waters with complicated bathymetry. Model grid construction is guided by the opposing necessities for resolving small-scale variations within convoluted coastlines and for reducing the computational load. Choice and tuning of subgrid parameterization schemes must accurately reflect known local conditions. Data assimilation schemes will vary depending on the volume and quality of real-time data available. Given the high density of satellite data, subsampling for model assimilation is necessary to reduce the computational load while reflecting the full range of the variable of interest. Similarly to color-coded depiction of satellite data, lower and upper data range boundaries are chosen to reflect known natural local variability so as to produce informative graphics with high information content. If time series graphs of model output for specific locations is contemplated (virtual buoys), numerical procedures for data extraction at these points must be implemented. Further detail regarding model implementation may be found in Chap. 6 of this book.

8.3 Deployment and Maintenance of Ocean Observing Platforms in the Coastal Zone

The wide range of mobile and fixed platforms now available for ocean observing requires specialized skills and equipment for each task. Significant expense may be incurred in platform deployment, maintenance, and retrieval. Three examples, those of coastal buoys, HF radar, and gliders, are discussed below.

8.3.1 Buoy Deployment and Maintenance

Design of instrument-bearing buoy systems is discussed elsewhere in this book (Chap. 3, Sect. 3.2). Information regarding buoy system design and engineering has been compiled by Snyder (1969). Intermediate size coastal ODAS buoys require heavy mooring anchors with weight in excess of the buoy displacement weight (around one metric ton) by at least 30%. Transportation and deployment of such buoys is preferentially accomplished aboard specialized buoy-tending vessel equipped with purpose-designed buoy handling gear and low freeboard mid sections or stern ramps to facilitate operations. In their absence, oceanographic vessels or other vessels or barges equipped with heavy lifting gear may be used. When deployed from a vessel, the buoy is released first. Once adrift, the mooring tackle is paid out as the vessel progresses away from the buoy. The anchor is weighed off deck and released into the water at the designated station before the final lengths of tackle are paid out (Fig. 8.2).

An alternate means of buoy system deployment is for anchor and mooring line to be floated at dockside using heavy-duty pneumatic lifting bags. Dockside cranes facilitate transfer of buoy and mooring tackle to the water. The entire buoy system may be then towed to the deployment site. Exact buoy deployment site is recorded at the time of anchor release.

Fig. 8.2 Buoy-first deployment; buoy adrift and anchor on deck as tackle is paid out

ODAS buoy retrieval for instrument and platform maintenance may entail recovery of the entire mooring, essentially the reverse of the deployment operation. However, under favorable conditions of depth, visibility, and bottom nature, divers may have access to the anchor end-tackle, making anchor retrieval unnecessary thus facilitating the operation. Maintenance of coastal buoys equipped with chain moorings is commonly performed at yearly intervals to allow for chain and instrument replacement or refurbishing. Maintenance may be more frequent in waters of high biological activity due to loss of instrument performance attendant to biofouling. In all cases, instruments are dismounted for biofouling removal (if submerged), cleaning inspection and on-site maintenance or shipment to manufacturers for recalibration and refurbishing (if necessary). As mentioned above, best practice is to replace the old instrument with a freshly calibrated unit. Immediately upon recovery, the buoy hull, stand, and other submerged surfaces must be freed of biofouling by mechanical means in anticipation of the onset of putrefaction (especially rapid in tropical climates). All buoy-specific equipment is inspected and repaired or replaced. Electrical bulkhead fittings are decoupled, cleaned, lubricated if required, and reinstalled. Structural repairs are performed as required, batteries are replaced (if past their operational life), buoys are repainted with anti-fouling paint and new wire rope and chain-and-tackle, incorporating in-line steel instrument cages, is prepared and attached. Following instrument installation, and verification of instrument and communications system performance, the buoy system is ready for redeployment in days to weeks.

Specialized moorings such as wave buoys and the MapCO2 design may use two-point moorings incorporating elastic elements in the mooring design in addition to the normal chain, steel rope, or synthetic fiber cable (see Fig. 3.4). Tensioning equipment is required for these operations in order to pre-set tension of the elastic mooring. The MapCO2 mooring, designed for fore-reef deployment on a coral reef in a region of low tidal amplitude specifies a fully tensioned mooring. The wave buoy for insular shelf deployment may specify a tensioned elastic mooring and a second inelastic mooring incorporating a double catenary design to allow for buoy heave.

8.3.2 High Frequency Radar Deployment and Maintenance

High frequency (HF) radar antennae are deployed as close to the land-sea interface as practicable to maximize seawater conductivity for signal transmission. For older units equipped with separate Tx and Rx antennae care must be taken to allow minimum separation between the antenna pair as per manufacturer specifications. Excessive length of cable runs to electronic control equipment will also impair performance. A few loops of the cable about one foot in diameter are normally turned and affixed to the antennae to reduce radio frequency interference. Clamp-on ferromagnetic filters may also be affixed to the cable at the console bulkhead fitting to further reduce interference. Temporary antenna emplacements require only a simple base (a parasol base placed upon a plywood sheet will suffice) and stabilizing guy wires affixed to stakes. More permanent emplacements usually feature a concrete base with a fabricated aluminum swivel-type antenna base mounted on J-bolts embedded in the concrete base to simplify lowering of the antenna for servicing (Fig 8.3).

Antennae Tx and Rx control modules and computer interface must be contained in low-humidity climate controlled housing either within previously constructed buildings or in commercial outdoor instrument housing units. These shelters are furnished with adjustable shelving and air conditioning. GPS antennae for system synchronization may be mounted atop the housing. Internet connectivity for data transmission may be wired, through cellular network connection, directional microwave or satellite communication systems.

Normally, HF radar equipment requires little maintenance since there are no moving parts to wear. Nevertheless, connector corrosion in the nearshore environment may require periodic cleaning or replacement of bulkhead and in-line connectors. Guy wires supporting antennae upright in provisional deployments may sag and should be checked regularly. It is prudent to dismount and store the antennae

Fig. 8.3 Antenna swivel base for permanent HF radar emplacement

securely in anticipation of stormy weather when wind and lightning may pose hazards to the equipment although, in the latter case, antennae may be provided with lighting protection devices.

8.3.3 Glider Deployment and Recovery

Glider preparation for deployment requires knowledge of seawater density structure in the deployment area in order to adjust vehicle buoyancy to dive to the programmed depth and ascend again to the surface. Deployment can be effected from a small boat by two operators. The glider, mounted on its carrier dolly, can be lifted to the gunwales and slipped off the dolly with minimum effort. Recovery aboard a small, low freeboard vessel can be easily performed by hand. Some gliders are equipped with faired hand grips for this purpose. Gliders may be deployed from larger ships by lowering the vehicle into the water with a crane or davit suspended in fabric slings. Recovery aboard large vessels is complicated because of the hazard of collision with the larger vessel. Custom designed nets or webbing slings may be used. A simple often-used alternative, when sea state allows, is to lower a small boat for glider recovery.

Table 8.1 CariCOOS partnerships

Organization	System	Services
University of Maine, Physical Oceanography Group	ODAS buoys	Data handling, quality assurance/quality control, buoy maintenance, capacity building
Commercial Divers Inc.	ODAS buoys	Buoy deployment/recovery
Rutgers University, Coastal Ocean Observing Laboratory	HF radar	Data handling, quality assurance/quality control, system maintenance, capacity building
National Oceanic and Atmospheric Administration, Atlantic Oceanographic and Meteorological Laboratory	Gliders, drifters	Data handling, quality assurance/quality control, capacity building
National Oceanic and Atmospheric Administration, Pacific Marine Environmental Laboratory	MAPCO2 buoy	Data handling, quality assurance/quality control, instrument maintenance, capacity building
University of California San Diego, Coastal Ocean Data Information Program	Wave buoy	Data handling, quality assurance/quality control
WeatherFlow Inc.	Meteorological stations	Data handling, quality assurance/quality control, instrument maintenance
University of South Florida	Satellite imagery	Data reception, data processing, image rendering
European Space Agency	Satellite imagery	Data reception, data processing, image rendering

8.4 Partnerships in Coastal Ocean Observing

Developing expertise in operation and maintenance of the wide range of instruments and platforms, in implementation and depiction of numerical on ocean models, and in extraction and depiction of satellite ocean color data requires a large, well-trained work force (and a budget to match) which can only be achieved with generous long-term, financial and institutional support. Nascent ICOOSs consequently find it useful to develop partnerships with more mature organizations specializing in one or more observing components to aid in system development, implementation operation and maintenance, as well as human resources capacity building. Partnerships established by CariCOOS are described in Table 8.1.

8.5 Applied and Scientific Research in Coastal Ocean Observing

Scientists occupy a unique position as ICOOS stakeholders. Given the specialized knowledge required to successfully operate such a system, oftentimes scientists are system operators as well as stakeholders. Climate science in particular finds great use for ICOOS time series data since it is concerned with short-term climatic anomalies as well as long-term climate change. Since ecosystem health is itself tied to climate variability, this field also makes use of such data. Applied science can also find fruitful turf for engagement in ocean observing. In many cases, collaboration to implement so-called *dual-use technologies* can be mutually beneficial. Four examples of successful scientific research supported by CariCOOS are related below.

Building upon a successful project designed to test autonomous pH and pCO_2 instruments (Grey et al. 2012), CariCOOS engaged to assist in the operation and maintenance of a MAPCO2 buoy emplaced on the fore reef of a coral reef off the coast of Puerto Rico in a collaborative agreement with NOAA PMEL. In addition to onboard air and water pCO_2 measurement equipment, the buoy is equipped with a CTD, a dye-based pH sensor and satellite communications. CariCOOS is tasked with yearly buoy recovery and maintenance and with initialization and validation of instruments and communication equipment performance prior to redeployment. CariCOOS moreover engages in weekly sampling for precise laboratory analysis of carbonate chemistry (carbonate alkalinity and pH). These instrumental records and laboratory analyses have established the remarkably wide variability of carbonate chemistry on this coral reef, a pCO_2 range, for example, of slight sub-saturation from below, the current mean atmospheric content of 405 ppm, to supersaturation above 500 ppm. This large variability has however impeded unequivocal discernment of the well-known global increase trend of about 5 ppm per year.

Security concerns regarding illegal vessel traffic across the Mona Passage between the islands of Puerto Rico and Hispaniola allowed support of a concerted effort to explore the use of HF radar as a dual-use technology to, in addition to detecting surface currents, detect and track vessels in transit within this passage between the northeastern Caribbean Sea and the western tropical Atlantic Ocean (WTA), an example of applied research. Given the complementary interest of CariCOOS in acquiring the capability for surface current measurements in the region, a collaborative effort with the National Center for Secure and Resilient Maritime Commerce and Coastal Environments (CSR) allowed emplacement and operation of two mid-range HF radar sites off the west coast of Puerto Rico. Experimental measurements confirmed that large vessels could be reliably tracked in this environment (as had been shown for other environments), but smaller vessels with lower superstructure were, on the contrary, not detectable due to the comparatively long wave amplitude of the vertically polarized probe signal relative to vessel size. More to the point of ocean observing, operational HF radar coverage of surface currents in the eastern Mona Passage has been available since 2009 as a result of these efforts (Corredor et al. 2011). After 2015, CariCOOS assumed full operational control of these pioneering installations and entered into an agreement with the Coastal Ocean Observing Laboratory of Rutgers, the State University of New Jersey (RU-COOL) for joint installation, operation, and maintenance of three long range emplacements expanding the network coverage to include the southern Caribbean coast of Puerto Rico. As for all other HF radar emplacements monitoring US coastal waters, data is initially received, processed, disseminated, and archived in local servers, and subsequently retrieved from these servers and reprocessed, disseminated, and archived by the Coastal Observing Research and Development Center (CORDC) at the University of California San Diego and by NOAA.

The geographic location of Puerto Rico and the Virgin Islands makes them particularly vulnerable to the impact of Atlantic Hurricanes, but prediction of hurricane intensity is hampered by lack of real-time data concerning upper ocean heat content, the driver of cyclone intensification. Accordingly, in 2015, CariCOOS entered into an agreement with the NOAA Atlantic Oceanographic and Meteorological Laboratory to deploy ocean gliders performing transects north and south of Puerto Rico with the object of generating detailed data on upper water mass structure in the path of oncoming hurricanes and in their wake for assimilation into coupled ocean-atmosphere models. Results reported by Goni et al. (2017) revealed a 50% improvement in wind intensity forecasts for Hurricane Gonzalo traversing the WTA with data assimilation from the glider there deployed. These results strengthen the argument for deployment of an expanded, dedicated, operational network of gliders in the region to improve hurricane intensity forecasts.

Another early effort in the collaboration with the Mid-Atlantic Regional Coastal Ocean Observing System to explore occupation of the Caribbean Time Series Station (CaTS – a precursor to CariCOOS) by means of autonomous vehicles allowed the discovery of a train of 40 m amplitude internal waves exiting the Mona Passage following the deployment of a glider controlled and operated at RU COOL (Corredor 2008). Reassuringly, the glider, deployed days earlier, was observed to surface and dive again as planned alongside a research vessel concurrently occupying CaTS in deep offshore water.

References

Corredor JE. Development and propagation of internal waves in the Mona Passage. Sea Tech. 2008;49(10):48–50.

Corredor JE, Amador A, Canals M, Rivera S, Capella JE, Morell JM, Glenn S, Handel E, Rivera E, Roarty H. Optimizing and validating high frequency radar surface current measurements in the Mona passage. Mar Technol Soc J. 2011;45(3):49–58.

Goni GJ, Todd RE, Jayne SR, Halliwell G, Glenn S, Dong J, Curry R, Domingues R, Bringas F, Centurioni L, DiMarco SF, Miles T, Morell J, Pomales L, Kim H-S, Robbins PE, Gawarkiewicz GG, Wilkins J, Heiderich J, Baltes B, Clone JJ, Seroka G, Knee K, Sanabria RR. Autonomous and Lagrangian Ocean observations for Atlantic tropical cyclone studies and forecasts. Oceanography. 2017;30(2):92–103. https://doi.org/10.5670/oceanog.2017.227.

Grey SEC, De Grandpre MD, Langdon C, Corredor JE. Short-term and seasonal pH, pCO2 and saturation state variability in a coral-reef ecosystem. Global Biogeochem Cycles. 2012;26:GB3012. https://doi.org/10.1029/2011GB004114.

Snyder RM. Buoys and buoy bystems. In: Myers JJ, Holm CH, McAllister RF, editors. Handbook of ocean and underwater engineering. New York: Copyright by North American Rockwell Corporation, McGraw-Hill; 1969. 1969; 9-81-9-115.

Statsewich H, Weingartner T, Grunau B, Egan G, Timm J. A high-latitude modular autonomous power, control and communication system for application to high-frequency surface current mapping radars. Mar Tech Soc J. 2011;45(3):59–68.

Whelan C, Barrick D, Lilleboe PM, Breivik Andrés PMO Kjelaas A, Alonso-Martirena A. Rapid deployable HF RADAR for Norwegian emergency spill operations. OCEANS 2010 IEEE-Sydney; 2010. https://doi.org/10.1109/OCEANSSYD.2010.5603848. https://www.researchgate.net/publication/233918950_Rapid_deployable_HF_RADAR_for_Norwegian_emergency_spill_operations (Accessed 1/20/2018).

Correction to: Coastal Ocean Observing

Jorge E. Corredor

Correction to:
J. E. Corredor, *Coastal Ocean Observing*,
https://doi.org/10.1007/978-3-319-78352-9

The published version of this book had missed the grey box rendering on multiple paragraph headings throughout the book and the author affiliation was inadvertently published with the incorrect city name. Also there were typo error in pages 44, 78, 80, 108 and 113. These rendering have been corrected and the affiliation has been updated throughout the book.

The updated online version of this book can be found at
https://doi.org/10.1007/978-3-319-78352-9

J. E. Corredor, *Coastal Ocean Observing*,
https://doi.org/10.1007/978-3-319-78352-9_9

Afterword

A short history of early steps taken to implement instrumental coastal ocean observing using *electronic* means provides perspective from which to judge the monumental advances achieved in the last century. Demand for radio telecommunications early in the last century fueled the development of electronic vacuum-tube diodes providing means to rectify AC current and triodes that could amplify tenuous electromagnetic radio signals detected by radio antennae. Radio antennae thus in effect constituted the first widespread commercially available electronic *sensors*. The transistor invented in 1947, a solid-state device that outperformed vacuum tubes in ruggedness, energy economy, switching speed, and bulk allowed for the development of a wide range of electronic instrument and their miniaturization.

Perusal of texts published between 1947 and the 1970s provide a view of developments in the field. *The Oceans* (Sverdrup et al. 1942), a widely admired treatise of oceanography, provides a few nuggets of information regarding the state of electrical instrument measurements at that date. The authors quote advances in *recent years* of photoelectric cells (resistance-based cadmium sulfide optical sensors), precursors to modern photodiodes for the study of seawater optical attenuation properties. Interestingly, the authors mention *photronic cells* but alas provide no further reference to their nature or operation. Analog-driven quartz crystal acoustic transducers were then already available, and various types of electromechanical current meters are mentioned providing dial displays or paper tape recordings. Such instruments made use of Wheatstone bridges to assess current differentials, but power amplification of such devices was restricted to vacuum-tube-equipped devices.

Fifteen years later, Von Arx (1962) in his *An Introduction to Physical Oceanography* mentions F. Wener's 1922 design of a shipboard conductivity bridge for assessing salinity used aboard coast guard vessels on the International Ice Patrol. Von Arx further describes current meters of *more recent design* that *favor electrical cable both as the means of suspending the instrument and for*

J. E. Corredor, *Coastal Ocean Observing*,
https://doi.org/10.1007/978-3-319-78352-9

transmitting the information to the ship's deck, to floating radio transmitters, or to shore stations using a ducted propeller powering an electromagnetic induction circuit of the author's design. Another electronic current measurement instrument, the geomagnetic electrokinetograph, of Von Arx's design is mentioned in Pickard's 1963 *Descriptive Physical Oceanography*. Only 7 years later, Myers, Holm, and McAllister (1969) in their definitive (for the times) multiple-author *Handbook of Ocean and Underwater Engineering* describe many commercially available electronic oceanographic measurement instrumentation available at the time but mostly for short-term industrial applications. Finally, at the threshold of the digital revolution, Williams (1973) in his *Oceanographic Instrumentation* describes and depicts more capable submersible field models of instruments for the measurement of temperature (XBT thermistors), salinity (induction conductivity), light irradiance, and transmission and scattering (resistance photocells). Already wave height could be measured by differential pressure using paired strain gauge transducers, and a variety of electronic current measurement devices was available. Electrochemical analyses were limited to Clark-type electrodes for dissolved oxygen measurement and glass membrane pH sensors. Most of these sensors relied on analog technology making data recording laborious and imprecise, and only a few incorporated crude digital technology using perforated cardboard or photographic film.

As described in this book, since then, the availability of electronic environmental measurement devices has grown exponentially. These advances are in large part due to the advent of electronic digital data processing, archiving, and transmission made possible by continued development of more capable electronic devices and of miniaturization and incorporation of such devices into integrated circuits, coupled to advances in electronic navigation, communications, and numerical simulation. Many instruments based on different detection principles are now available to measure a single variable, tuned at times to a specific range or environmental application in compact durable formats capable of extended autonomous deployment. Platforms, then mostly limited to ships and coastal installations, multiplied to encompass the oceans from outer space to the sea bottom. Digital numerical simulations, then inexistent, now provide model forecast guidance for diverse applications including navigation, recreation, search and rescue, and spill response.

Sustained maintenance and further development of operational coastal ocean observing systems will depend on their yield to societal benefit and on proper recognition of the benefit thus accrued. Science and technology will undoubtedly be among the beneficiaries yielding in turn more capable sensors, platforms, and numerical models and, thus, more capable observing systems. Capacity for sound governance and fiscal integrity will determine the outcome.

References

Myers JJ, Holm CH, McAllister RF. Handbook of ocean and underwater engineering. NY: Copyright by North American Rockwell Corporation: McGraw-Hill; 1969. p. 1094.
Pickard GL. Descriptive physical oceanography. Oxford: Pergamon Press; 1963. p. 200.
Sverdrup HU, Johnson MW, Fleming RH. The oceans, their physics, chemistry, and general biology. New York: Prentice-Hall; 1942. p. 1087.
Von Arx WS. An introduction to physical oceanography. Reading: Addison-Wesley Publishing Company Inc.; 1962. 422 pp
Williams J. Oceanographic instrumentation. Annapolis: Naval Institute Press; 1973. 189 pp

Index

J. E. Corredor, *Coastal Ocean Observing*,
https://doi.org/10.1007/978-3-319-78352-9

Printed in the United States
By Bookmasters